T0391271

SOLUTIONS AND X-RAY NON-3D PHASE STRUCTURE ANALYSIS OF SOFT MATTER

SOLUTIONS AND X-RAY NON-3D PHASE STRUCTURE ANALYSIS OF SOFT MATTER

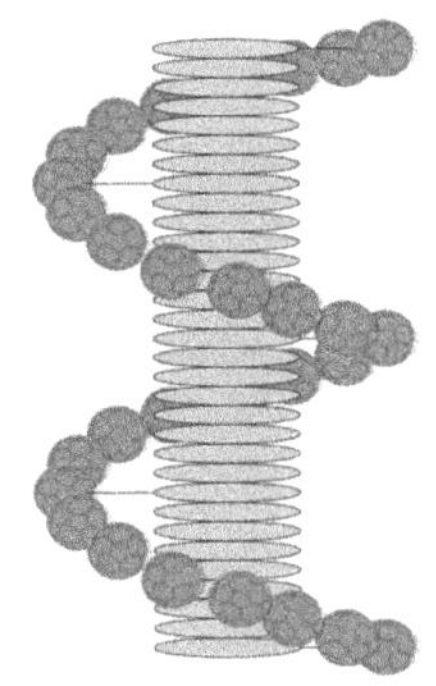

Kazuchika Ohta
Shinshu University, Japan

World Scientific
NEW JERSEY • LONDON • SINGAPORE • BEIJING • SHANGHAI • HONG KONG • TAIPEI • CHENNAI • TOKYO

Published by

World Scientific Publishing Co. Pte. Ltd.
5 Toh Tuck Link, Singapore 596224
USA office: 27 Warren Street, Suite 401-402, Hackensack, NJ 07601
UK office: 57 Shelton Street, Covent Garden, London WC2H 9HE

Library of Congress Control Number: 2023036615

British Library Cataloguing-in-Publication Data
A catalogue record for this book is available from the British Library.

SOLUTIONS AND X-RAY NON-3D PHASE STRUCTURE ANALYSIS OF SOFT MATTER

ISBN 978-981-127-241-7 (hardcover)
ISBN 978-981-127-242-4 (ebook for institutions)
ISBN 978-981-127-243-1 (ebook for individuals)

For any available supplementary material, please visit
https://www.worldscientific.com/worldscibooks/10.1142/13310#t=suppl

Typeset by Stallion Press
Email: enquiries@stallionpress.com

Preface

Speaking of X-ray structure analysis, most people would think of crystal structure analysis. A crystal is a state in which atoms or molecules are regularly arranged in three-dimensional space. However, some substances do not have 3D structures, but have 1D, 2D, 1D$\oplus$1D, 2D$\oplus$1D or 1D$\oplus$1D$\oplus$1D structures. Here the symbol $\oplus$ stands for direct sum in linear algebra and is different from ordinary sum $+$. For example, 2D$\oplus$1D means the direct sum of a 2D subspace and an orthogonal 1D subspace, not 3D. In my previous book *Physics and Chemistry of Molecular Assemblies*, which was published in April 2020, I showed for the first time that molecular assemblies can be represented in a unified manner from such subspaces of linear algebra, and I wrote the theory and method of X-ray structure analysis of the liquid crystal phase structures with the dimensional structure of 1D, 2D, 1D$\oplus$1D, 2D$\oplus$1D, and 1D$\oplus$1D$\oplus$1D. Furthermore, the X-ray liquid crystal structure analysis methods of 11 representative examples were concretely described. In addition, 110 X-ray data were posted as the end-of-chapter problems in Chapter 3. Using the methods which I developed, these 110 problems can be solved in principle with a calculator, graph paper, a ruler, and a compass, although it takes time.

However, when you use only a calculator, a graph paper, a ruler and a compass, it would probably take more than one year to analyze all the observed X-ray data in Table 13. Therefore, our laboratory research group has developed and polished up a computor program named as "Bunseki-kun." By using this program, we can easily analyze the liquid crystal phase structures from the observed d spacings. Thus, when you use this program, you will be able to analyze the liquid crystal phase and the other soft matter phase structures within 1/100th to 1/1000th of the time. For those who purchase this book, you will be able to freely download this program by following the instructions given in the Supplementary Material.

X-ray structural analysis of 3D crystals is a well-established discipline and technique. However, the X-ray structural analysis of non-3D soft matter described above has not yet been established as an academic field and technique. So, for example, when the observed spacing values are in ratios like $d_1 : d_2 : d_3 : \cdots = 1 : \frac{1}{\sqrt{3}} : \frac{1}{2} : \cdots$, the phase structure was analyzed from this ratio to have a 2D hexagonal lattice. If the number of observed spacing values increases, the identification becomes more difficult, and unfortunately there are many mistakes of the indexation and identification in the reports. Utilization of our program "**Bunseki-kun Ver. 3**" eliminates such mistakes and makes indexation very easy. In my previous book, in order to make this method and technique easy to understand, I have explained, firstly, the outline of the 3D crystal structure analysis method in Chapter 1, followed by crystalline polymorphism in Chapter 2. Finally in Chapter 3, I have explained the liquid crystalline polymorphism and described a new X-ray liquid crystal structure analysis method of "**Reciprocal Lattice Method.**"

This approach is by no means limited to liquid crystal structures. In fact, it can be used for soft matter in general. Therefore, in this book, I have added six more problems to the previous 110 problems. In the additional problems, block copolymer and spider silk are included. This is because soft matter includes liquid crystals, micelles, concentrated solutions of biopolymers, block copolymers, and fiber structures such as silk and spider silk. Until now, there have been no theories and techniques for X-ray structural analysis that took a colligative viewpoint for general soft matter, and many fields are developing independently without co-operation from other fields. Therefore, as described the viewpoint of linear algebra in my previous book, we can see all the soft matters as the direct sum of subspaces. So, it is possible to analyze the phase structures of wide-range of soft matter from my developed new methods: both "Reciprocal Lattice Method" and "FlexiLattice Method" which are published for the first time in this book.

Part I of this book provides the solutions to the end-of-chapter problems in my previous book, and Part II describes how to use the non-3D soft matter X-ray structural analysis program, **Bunseki-kun Ver.3**. I hope that this program will be used not only for the exercises but also for your actual research.

You may freely distribute and share this computer program with colleagues and students in your laboratory for the purpose of promoting your research. However, please refrain from transferring it to a third party or selling it with a different name for commercial purposes. If you can improve this program to something better in the future, please publish it with the original publisher, World Scientific Publishing. In order to improve the academic level of this field, we should collaborate with all the soft matter researchers around the world to improve this program. This is the same idea as the free OS, Linux. I am no longer young (71 years old) to keep improving this program, so I wish many young brilliant researchers around the world to improve this program instead of me and develop this fascinating research field further.

Kazuchika Ohta
Emeritus Professor, Shinshu University
June 2023

Contents

Part I

Solutions to End-of-Chapter Problems in “Physics and Chemistry of Molecular Assemblies”

Chapter 1

Solutions to End-of-Chapter Problems

Problem 1. Explain what white X-rays and characteristic X-rays are.

Both white X-rays and characteristic X-rays are electromagnetic waves generated by collision of high-speed electrons on a target metal. The difference between these X-rays is as follows.

White X-rays: it is also called continuous X-rays. Since the electrons collide on the target metal to be decelerated, the bremsstrahlung occurs to emit the X-rays. As shown in Figure 1, a continuous spectrum having a wide wavelength can be observed and the spectrum changes depending on the applied voltage. The wavelength of continuous X-rays is equal irrespective of the type of target metal and depends only on the applied voltage.

Characteristic X-ray: An electromagnetic wave having a wavelength specific to the type of the target metal is generated. As shown in Figure 2, a sharp peak appears at the wavelength characteristic to the target metal. This is attributable to the different energy levels depending on the target metal.

Problem 2. Geometrically prove Bragg's reflection condition.

Figure 3 shows a schematic representation of a crystal in which atoms and molecules are arranged regularly and periodically.

When a crystal having such a periodic structure is irradiated with X-rays, diffraction occurs. As shown in this figure, it is assumed that an X-ray hits the atoms (or molecules) in the first and second layers

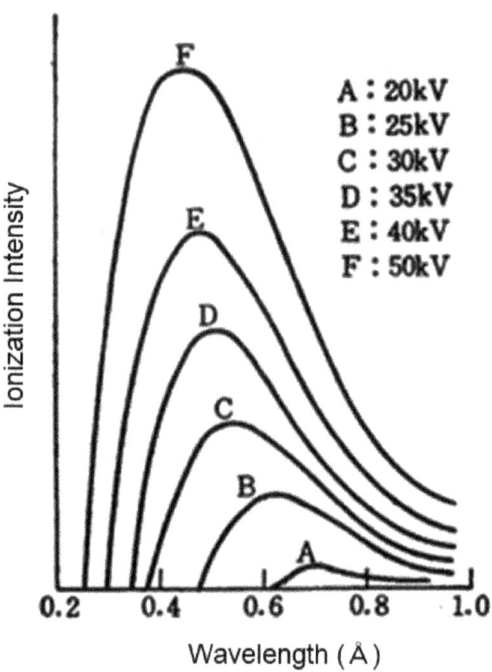

Figure 1. Dependence of ionizition intensity and wavelength on the applied voltage for the generation of X-ray.

of the crystal at an incident angle θ and then it is reflected by the atoms. When compared to the reflection from the first layer with the reflection from the second layer, the reflection from the second layer has an extra path difference of $2d\sin\theta$. When this path difference includes an integer (n) times of the waves, these two reflected X-rays are overlapped in phase to appear as dark lines on the photographic plate. On the other hand, when the path difference includes half-integer ($n + 1/2$) times of the waves, these two reflected X-rays are exactly reversed in phase to cancel each other; nothing appears on the photographic plate. Therefore, the condition under which a dark line appears is

$$2d\sin\theta = n\lambda,\ n = 1, 2, 3 \ldots\ (d\text{: surface spacing; } \theta\text{: incident angle; } \lambda\text{: X-ray wavelength}) \tag{1}$$

This is called the Bragg condition.

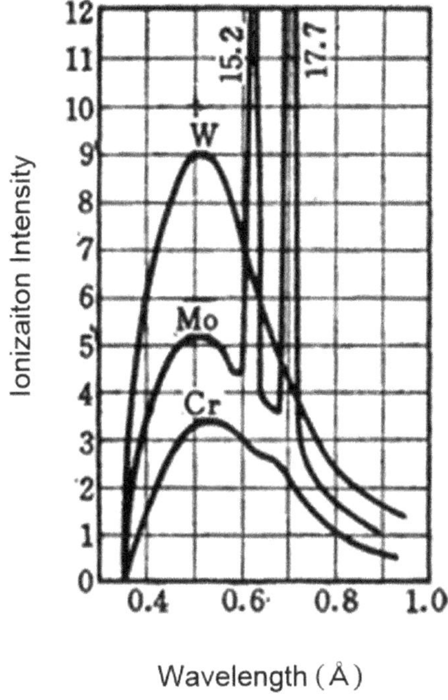

Figure 2. Emission of X-ray characteristic to the type of target metal.

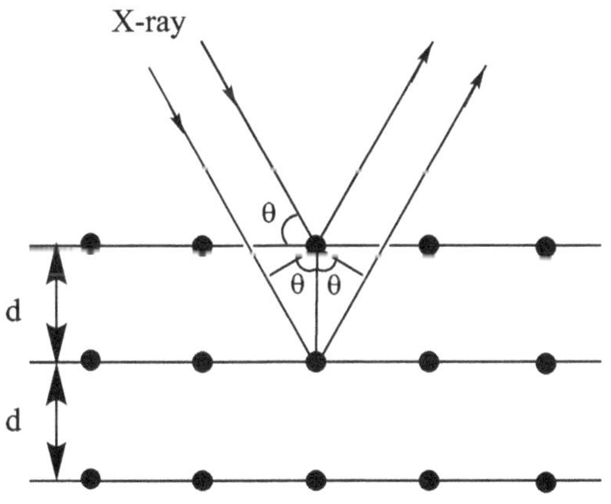

Figure 3. Bragg's law (1).

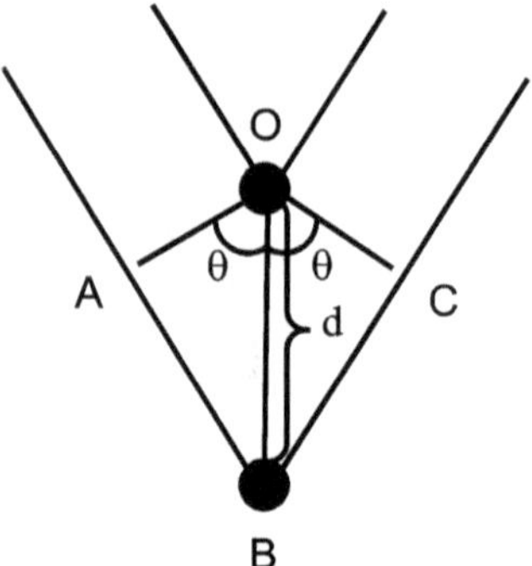

Figure 4. Bragg's law (2).

This phenomenon of X-ray diffraction was discovered by the Bragg family, who were awarded the Nobel Prize in Physics in 1915.

The detailed geometrical proof is as follows. Figure 4 shows an enlarged version of Figure 3.

In this figure, the length of the side AB of △OAB can be obtained from the law of sines

$$\frac{d}{\sin 90} = \frac{AB}{\sin\theta}$$
$$AB = d\sin\theta$$

From △OAB ≡ △OCB, AB = CB

Therefore, the thick line (path difference) in Figure 4 can be calculated as

$$2\text{AB} = 2d\text{sin}\theta = \text{(path difference)}$$

As described above, when the path difference is an integral multiple, the Equation (1) of Bragg's reflection condition holds and strong diffraction is observed.

Thus, the reflection condition of Bragg was geometrically proved.

Problem 3. In an orthorhombic lattice, mark and show each plane of (110), (111), (210), (200), (310) represented by Miller index.

Firstly, the Miller index is converted to the Weiss index.

Miller index	Reciprocal	Weiss index
(1 1 0)	$\rightarrow$	$(1\ 1\ \infty)$
(1 1 1)	$\rightarrow$	$(1\ 1\ 1)$
(2 1 0)	$\rightarrow$	$(\frac{1}{2}\ 1\ \infty)$
(2 0 0)	$\rightarrow$	$(\frac{1}{2}\ \infty\ \infty)$
(3 1 0)	$\rightarrow$	$(\frac{1}{3}\ 1\ \infty)$

Next, the surface represented by these Weiss indices is shown in the figure below.

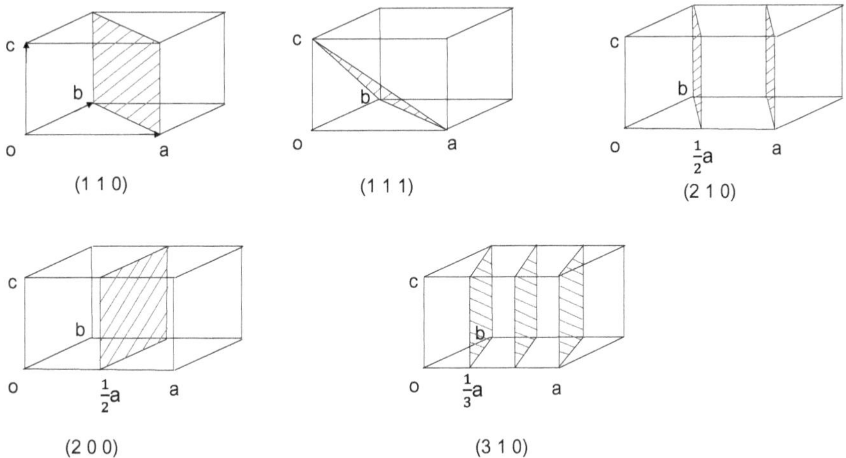

Problem 4. Calculate the spacing d_{hk0} of a two-dimensional rectangular lattice.

We consider the spacing d_{hk0} of a two-dimensional rectangular lattice as shown in Figure 5.

Then, we consider a straight line passing through two points $(\frac{a}{h}\ 0)$, $(0\ \frac{b}{k})$.

In general, the equation of a straight line passing through two points (A 0) and (0 B) in a xy plane can be described as

$$\frac{x}{A} + \frac{y}{B} = 1 \tag{2}$$

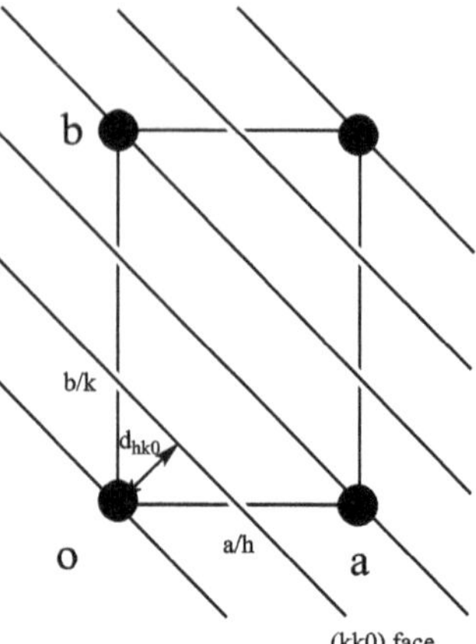

Figure 5. Relationship among a spacing d, Miller indices ($hk0$) and lattice constants (a and b) in the case of a 2D rectangular lattice.

Hence,

$$\frac{x}{\left(\frac{a}{h}\right)} + \frac{y}{\left(\frac{b}{k}\right)} = 1 \tag{2'}$$

$$\frac{b}{k}x + \frac{a}{h}y = \frac{ab}{hk}$$

$$\frac{b}{k}x + \frac{a}{h}y - \frac{ab}{hk} = 0 \tag{2''}$$

Also, in general, the distance d from the point $\mathrm{P_0}$ $(x_0\, y_0)$ to the straight line $ax + by + c = 0$ can be described as

$$\mathrm{d} = \frac{|ax_0 + by_0 + c|}{\sqrt{a^2 + b^2}} \tag{3}$$

When $(x_0\, y_0) = (0\ 0)$

$$\mathrm{d} = \frac{|c|}{\sqrt{a^2 + b^2}}$$

$$\therefore \frac{1}{d^2} = \frac{a^2 + b^2}{c^2} \tag{3'}$$

Therefore, the plane spacing d_{hk0} between the straight line (2″) and the origin O can be derived as

$$\frac{1}{d_{hk0}^2} = \frac{(\frac{b}{k})^2 + (\frac{a}{h})^2}{(\frac{ab}{hk})^2} = \frac{a^2k^2 + b^2h^2}{a^2b^2} = \frac{h^2}{a^2} + \frac{k^2}{b^2}$$

$$\therefore \quad d_{hk0} = \sqrt{\frac{1}{\frac{h^2}{a^2} + \frac{k^2}{b^2}}}$$

Problem 5. Calculate the spacing d_{hkl} of a three-dimensional orthorhombic lattice.

We consider the plane spacing $d_{hk\ell}$ of the three-dimensional oblique (orthorhombic) lattice as shown in Figure 6.

Then, we consider a straight line passing through three points $(\frac{a}{h}\,0\,0)$, $(0\,\frac{b}{k}\,0)$, $(0\,0\,\frac{c}{l})$.

In general, the equation of a straight line passing through three points (A 0 0), (0 B 0), (0 0 C) is

$$\frac{x}{A} + \frac{y}{B} + \frac{z}{C} = 1$$

which is transformed as

$$BCx + ACy + ABz - ABC = 0 \tag{4}$$

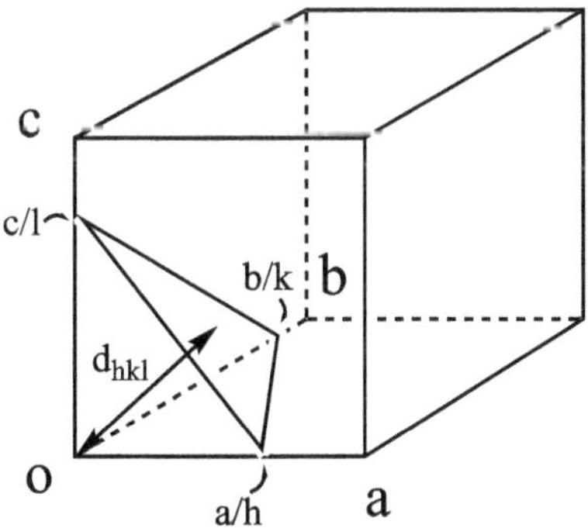

Figure 6. Relationship among a spacing *d*, Miller indices (*hkl*) and lattice constants (a, b and c) in the case of a 3D orthorhombic lattice.

Hence,

$$\frac{bc}{k\ell}x + \frac{ac}{h\ell}y + \frac{ab}{hk}z - \frac{abc}{hk\ell} = 0 \tag{4'}$$

Also, in general, the distance d from the point P_0 $(x_0\, y_0\, z_0)$ to the straight line $ax + by + cz + e = 0$ can be described as

$$\mathrm{d} = \frac{|ax_0 + by_0 + cz_0 + e|}{\sqrt{a^2 + b^2 + c^2}} \tag{5}$$

When $(x_0\, y_0\, z_0) = (0\,0\,0)$,

$$\mathrm{d} = \frac{|e|}{\sqrt{a^2 + b^2 + c^2}}$$

$$\therefore \quad \frac{1}{d^2} = \frac{a^2 + b^2 + c^2}{e^2} \tag{5'}$$

Therefore, the surface spacing $\mathrm{d}_{hk\ell}$ between the straight line (4′) and the origin O is

$$\frac{1}{d_{hk\ell}^2} = \frac{(\frac{bc}{k\ell})^2 + (\frac{ac}{h\ell})^2 + (\frac{ab}{hk})^2}{(-\frac{abc}{hk\ell})^2} = \frac{(bch)^2 + (ack)^2 + (ab\ell)^2}{(abc)^2}$$

$$= \frac{h^2}{a^2} + \frac{k^2}{b^2} + \frac{\ell^2}{c^2}$$

$$\therefore \quad d_{hk\ell} = \sqrt{\frac{1}{\frac{h^2}{a^2} + \frac{k^2}{b^2} + \frac{\ell^2}{c^2}}}$$

Problem 6. $SbCl_3$ is orthorhombic, and its lattice constants are a = 8.12 Å, b = 9.47 Å, c = 6.37 Å. Calculate the spacing of the (411) plane.

As given in the problem, $SbCl_3$ is orthorhombic, and its lattice constants are a = 8.12 Å, b = 9.47 Å, c = 6.37 Å. In this condition, the spacing of d_{411} can be derived as follows.

It is apparent from the solution for Problem 5 that, the spacing $d_{hk\ell}$ in an orthorhombic crystal is given by the following equation.

$$d_{hk\ell} = \sqrt{\frac{1}{\frac{h^2}{a^2} + \frac{k^2}{b^2} + \frac{\ell^2}{c^2}}}$$

Accordingly,

$$d_{411} = 1 \Big/ \sqrt{\frac{4^2}{8.12^2} + \frac{1^2}{9.47^2} + \frac{1^2}{6.37^2}}$$
$$\cong 1.89\,\text{Å}$$

Problem 7. Describe what you can reveal from single crystal X-ray diffraction and powder crystal X-ray diffraction, respectively.

In single crystal X-ray diffraction, it is possible to reveal both the three-dimensional structure of a molecule and the arrangement of the molecules in the crystal at the same time. In addition, you can obtain information about the distribution of outer shell electrons, chemical bonds, and electronic states. However, it can be said that it depends on the availability of a "good single crystal sample" in order to obtain highly accurate results in this analysis method. However, it is not easy to obtain a good single crystal.

On the other hand, in powder X-ray diffraction it is possible to reveal the arrangement of the molecules in the crystal, but impossible to know the three-dimensional molecular structure. The features of this analysis method are listed below.

- This method is widely available for all kinds of materials like as metal, alloy, inorganic substance, rock mineral, organic compound, polymer, biomaterial and so on. All solid samples can be measured non-destructively, and the measurement time is about several tens of minutes. It is possible to measure both crystalline and amorphous materials, and it is possible to measure not only powder samples but also plate-shaped and linear samples.
- Generally, substances show their own characteristic X-ray pattern. Therefore, you can identify the substance (qualitative analysis) in comparison with the data file of the standard substance.

- The diffraction pattern changes when the same compound has the different crystalline forms. Therefore, crystalline polymorphs can be identified from the diffraction patterns.
- It is possible to check the quality of crystallinity. The spread of the diffraction line can be measured to obtain an average particle size of 50 to 2000 Å. Amorphous materials give an amorphous diffraction pattern. The crystallinity of the polymer can also be measured.
- Crystal orientation can be investigated. It is used to study the structure of one-dimensional oriented samples such as fibers and wires, clay minerals, vapor-deposited films, and rolled materials.

Problem 8. Calculate the structure factor of NaCl, and obtain the extinction rule of the face-centered cubic lattice.

Figure 7 illustrates the unit cell of NaCl. The crystal structure factor is calculated as follows.

Crystal structure factor is given as

$$\mathrm{F(hk\ell)} = \sum_{i}^{N} f_i exp\{2\pi(hx_i + ky_i + \ell z_i)\} \tag{6}$$

N: Number of atoms (or molecules) in the unit cell
f_i: Scattering ability of the i-th atom (or molecule)

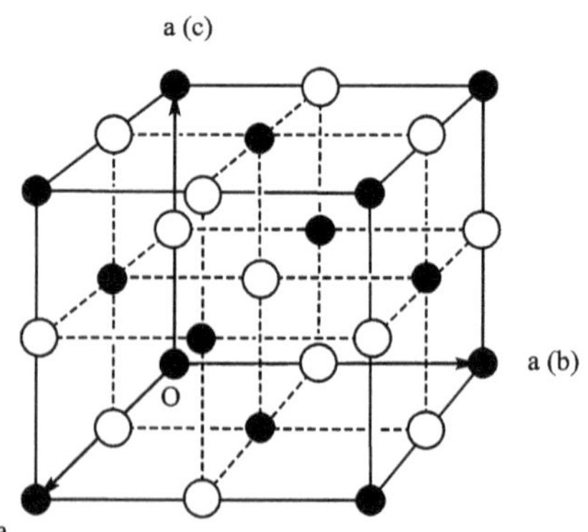

Figure 7. Unit cell of NaCl.

The scattering abilities of Na$^+$ and Cl^- are expressed as f_{Na} and f_{Cl}, respectively. Considering the ions at the positions specified in Figure 7, the scattering abilities are derived as Table 1 shown below.

For these calculations, the relationships, $\exp(i\pi) = -1$, $\exp(2\pi ih) = \exp(2\pi ik) = \exp(2\pi i\ell) = 1$, are used.

Therefore, the sum of total F(hkl) is

$$\begin{aligned} \mathrm{F(hkl)} &= f_{Na}[1 + (-1)^{h+k} + (-1)^{h+l} + (-1)^{k+l}] \\ &\quad + f_{Cl}[(-1)^h + (-1)^k + (-1)^l + (-1)^{h+k+l}] \end{aligned} \qquad (7)$$

Thus, the crystal structure factor of the unit cell of NaCl could be calculated.

From Equation (7), the following three results can be deduced.

(i) When all of h, k, and ℓ are even numbers, $\mathrm{F(hk\ell)} = 4(f_{Na} + f_{Cl}) \rightarrow$ High intensity
(ii) When all of h, k, and ℓ are odd numbers, $\mathrm{F(hk\ell)} = 4(f_{Na} - f_{Cl}) \rightarrow$ Low intensity
(iii) When one is odd and the other is even, or when one is even and the other is odd, $\mathrm{F(hk\ell)} = 0 \rightarrow$ disappears. From this condition, the extinction rule of fcc (h + k = odd number, k + ℓ = odd number, ℓ + h = odd number) can be derived.

Problem 9. Although both NaCl and KCl are face-centered cubic structures, why does KCl apparently look like a simple cubic structure in X-ray diffraction? (cf. Problem 8 and Moore Physical Chemistry p. 853)

In the NaCl salt, Na^+ ion has 10 electrons with a neon type of electronic structure, and Cl^- ion has 18 electrons with an argon type of electronic structure. So, Na^+ and Cl^- ions have different scattering abilities and a face-centered cubic lattice appears. On the other hand, in the KCl salt, both K^+ and Cl^- ions have 18 electrons with an argon-type of electron structure. So, their scattering abilities are the same and the difference between them disappears. Therefore, it looks like a simple cubic lattice. If you change all the filled circles to the open circles in Figure 7 above, you can see that it changes from a face-centered cubic lattice to a simple cubic lattice.

Table 1. Derivation of the crystal structure factor F(hkl) for a face-centered cubic lattice of NaCl.

Coordinate of Na	n in nf_{Na}	$\exp\{2\pi \mathrm{i}(\mathrm{h}x_i + ky_i + lz_i)\}$
(0 0 0)	$\frac{1}{8}$	$\exp\{0\} = 1$
(1 0 0)	$\frac{1}{8}$	$\exp\{2\pi \mathrm{i}(\mathrm{h})\} = 1$
(1 1 0)	$\frac{1}{8}$	$\exp\{2\pi \mathrm{i}(\mathrm{h} + k)\} = 1$
(0 1 0)	$\frac{1}{8}$	$\exp\{2\pi \mathrm{i}(k)\} = 1$
$\left(\frac{1}{2}\ \frac{1}{2}\ 0\right)$	$\frac{1}{2}$	$\exp\left\{2\pi \mathrm{i}\left(\frac{1}{2}\mathrm{h} + \frac{1}{2}k\right)\right\} = (-1)^{\mathrm{h}+k}$
$\left(\frac{1}{2}\ 0\ \frac{1}{2}\right)$	$\frac{1}{2}$	$\exp\left\{2\pi \mathrm{i}\left(\frac{1}{2}\mathrm{h} + \frac{1}{2}l\right)\right\} = (-1)^{\mathrm{h}+l}$
$\left(1\ \frac{1}{2}\ \frac{1}{2}\right)$	$\frac{1}{2}$	$\exp\left\{2\pi \mathrm{i}\left(\mathrm{h} + \frac{1}{2}k + \frac{1}{2}l\right)\right\} = (-1)^{k+l}$
$\left(\frac{1}{2}\ 1\ \frac{1}{2}\right)$	$\frac{1}{2}$	$\exp\left\{2\pi \mathrm{i}\left(\frac{1}{2}\mathrm{h} + k + \frac{1}{2}l\right)\right\} = (-1)^{\mathrm{h}+l}$
$\left(0\ \frac{1}{2}\ \frac{1}{2}\right)$	$\frac{1}{2}$	$\exp\left\{2\pi \mathrm{i}\left(\frac{1}{2}k + \frac{1}{2}l\right)\right\} = (-1)^{k+l}$
(0 0 1)	$\frac{1}{8}$	$\exp\{2\pi \mathrm{i}(l)\} = 1$
(1 0 1)	$\frac{1}{8}$	$\exp\{2\pi \mathrm{i}(\mathrm{h} + l)\} = 1$
(1 1 1)	$\frac{1}{8}$	$\exp\{2\pi \mathrm{i}(\mathrm{h} + k + l)\} = 1$
(0 1 1)	$\frac{1}{8}$	$\exp\{2\pi \mathrm{i}(k + l)\} = 1$
$\left(\frac{1}{2}\ \frac{1}{2}\ 1\right)$	$\frac{1}{2}$	$\exp\left\{2\pi \mathrm{i}\left(\frac{1}{2}\mathrm{h} + \frac{1}{2}k + l\right)\right\} = (-1)^{\mathrm{h}+k}$

$$\begin{aligned}\text{Sum} &= f_{Na}\left[\frac{1}{8} \times 8(1) + \frac{1}{2} \times 2(-1)^{h+k} + \frac{1}{2} \times 2(-1)^{h+l} + \frac{1}{2} \times 2(-1)^{k+l}\right]\\ &= f_{Na}\left[1 + (-1)^{h+k} + (-1)^{h+l} + (-1)^{k+l}\right]\end{aligned}$$

Table 1. (*Continued*)

Coordinate of Cl	n in nf_{Cl}	$\exp\{2\pi\text{i}(\text{h}x_i + ky_i + lz_i)\}$
$\left(\frac{1}{2}\ 0\ 0\right)$	$\frac{1}{4}$	$\exp\left\{2\pi\text{i}\left(\frac{1}{2}\text{h}\right)\right\} = (-1)^{\text{h}}$
$\left(1\ \frac{1}{2}\ 0\right)$	$\frac{1}{4}$	$\exp\left\{2\pi\text{i}\left(\text{h} + \frac{1}{2}k\right)\right\} = (-1)^{k}$
$\left(\frac{1}{2}\ 1\ 0\right)$	$\frac{1}{4}$	$\exp\left\{2\pi\text{i}\left(\frac{1}{2}\text{h} + k\right)\right\} = (-1)^{\text{h}}$
$\left(0\ \frac{1}{2}\ 0\right)$	$\frac{1}{4}$	$\exp\left\{2\pi\text{i}\left(\frac{1}{2}k\right)\right\} = (-1)^{k}$
$\left(0\ 0\ \frac{1}{2}\right)$	$\frac{1}{4}$	$\exp\left\{2\pi\text{i}\left(\frac{1}{2}l\right)\right\} = (-1)^{l}$
$\left(1\ 0\ \frac{1}{2}\right)$	$\frac{1}{4}$	$\exp\left\{2\pi\text{i}\left(\text{h} + \frac{1}{2}l\right)\right\} = (-1)^{l}$
$\left(1\ 1\ \frac{1}{2}\right)$	$\frac{1}{4}$	$\exp\left\{2\pi\text{i}\left(\text{h} + k + \frac{1}{2}l\right)\right\} = (-1)^{l}$
$\left(0\ 1\ \frac{1}{2}\right)$	$\frac{1}{4}$	$\exp\left\{2\pi\text{i}\left(k + \frac{1}{2}l\right)\right\} = (-1)^{l}$
$\left(\frac{1}{2}\ \frac{1}{2}\ \frac{1}{2}\right)$	1	$\exp\left\{2\pi\text{i}\left(\frac{1}{2}\text{h} + \frac{1}{2}k + \frac{1}{2}l\right)\right\} = (-1)^{\text{h}+k+l}$
$\left(\frac{1}{2}\ 0\ 1\right)$	$\frac{1}{4}$	$\exp\left\{2\pi\text{i}\left(\frac{1}{2}\text{h} + l\right)\right\} = (-1)^{\text{h}}$
$\left(1\ \frac{1}{2}\ 1\right)$	$\frac{1}{4}$	$\exp\left\{2\pi\text{i}\left(\text{h} + \frac{1}{2}k + l\right)\right\} = (-1)^{k}$
$\left(\frac{1}{2}\ 1\ 1\right)$	$\frac{1}{4}$	$\exp\left\{2\pi\text{i}\left(\frac{1}{2}\text{h} + k + l\right)\right\} = (-1)^{\text{h}}$
$\left(0\ \frac{1}{2}\ 1\right)$	$\frac{1}{4}$	$\exp\left\{2\pi\text{i}\left(\frac{1}{2}k + l\right)\right\} = (-1)^{k}$

$$\text{Sum} = f_{Cl}\left[\frac{1}{4} \times 4(-1)^{h} + \frac{1}{4} \times 4(-1)^{k} + \frac{1}{4} \times 4(-1)^{hl} + (-1)^{h+k+l}\right]$$

$$= f_{Cl}\left[(-1)^{h} + (-1)^{k} + (-1)^{l} + (-1)^{h+k+l}\right]$$

Problem 10. Carry out the crystal structure analysis of NaCl using the X-ray powder pattern shown in Figure 1–24 in the textbook. Follow Procedures from 1 to 5 written below. (cf. Section (1-13) in the textbook)

(Procedure 1) Using $\lambda = 1.5418\,\text{Å}$ of Cu Kα, calculate the interplanar spacing d of each diffraction line and the relative intensity ratio (I/I_1).

(Procedure 2) It is already known from microscopic observation that NaCl is cubic system. With this in mind, index each diffraction line.

(Procedure 3) Calculate the lattice constant of NaCl. If the indexation in Procedure 2 is correct, the value calculated from each diffraction line should be almost constant. Check your results in this way.

(Procedure 4) Determine the number of atoms in the unit cell using the observed density $d_{obs.} = 2.163\,\text{g·cm}^{-3}$ of NaCl.

(Procedure 5) Examine your indexing from the extinction rule derived in Procedure 2, and prove that the NaCl crystal has a face-centered cubic lattice.

(Procedure 1) First, you read all the angles (2θ) from the diffraction peaks and summarize the data like in Table 2.

Next, you calculate the spacings of d values by using Bragg's reflection condition (Equation 1).

Table 2. X-ray diffraction data observed for the powder of NaCl.

Peak No.	2θ
1	27.35
2	28.57
3	30.66
4	31.71
5	45.44
6	53.86
7	56.48

Table 3. X-ray diffraction data observed for the powder of NaCl.

Peak No.	Observed 2θ	d(Å)
1	27.35	3.261
2	28.57	3.124
3	30.66	2.916
4	31.71	2.822
5	45.44	1.996
6	53.86	1.702
7	56.48	1.629

Table 4. The relative intensities of the diffraction peaks.

Peak No.	Observed intensity (I)	Relative intensity ($I/I_1 \times 100$)
1	0.25	1.69
2	0.05	0.34
3	0.06	0.41
4	14.79	100.00
5	1.91	12.90
6	0.06	0.41
7	0.38	2.57

When $n = 1$ and $\lambda = 1.542$ Å (Cu Kα line) in $2d\sin\theta = n\lambda$ (order of reflection $n = 1, 2, 3, \ldots$),

$$d = \frac{1.542}{2\sin\theta}$$

Furthermore, each of the intensities of the diffraction peaks was read, and obtained the relative intensity ratio (I/I_1) when the intensity (I_1) of the strongest peak was assumed as 100.

(Procedure 2) If each of the $1/d^2$ values can be a simple integer ratio, each of the peaks can be indexed. Accordingly, each of the values $1/d^2 \times 100$ are calculated from the original d values as shown in Table 5. Next, each of the values of $1/d^2 \times 100$ is divided by the

Table 5. Calculations of the simplest ratios and the indexation by using these ratios.

Peak No.	d(Å)	$\frac{1}{d^2} \times 100$	Ratio 1	Ratio 2	(hkl)	$a = d\sqrt{h^2+k^2+l^2}$
1	3.260	9.41	1.000	3.00	(1 1 1)	5.65
2	3.124	10.25	1.089	3.26 ×	–	–
3	2.916	11.76	1.250	3.75 ×	–	–
4	2.822	12.56	1.335	4.00	(2 0 0)	5.64
5	1.996	25.10	2.667	8.00	(2 2 0)	5.65
6	1.702	34.52	3.668	11.00	(3 1 1)	5.64
7	1.629	37.68	4.000	12.00	(2 2 2)	5.64

(Arrows in the $\frac{1}{d^2} \times 100$ column: ×4 from 9.41 to 37.68; ×2 from 12.56 to 25.10; ×3 from 12.56 to 37.68.)

The simplest ratios — Indexation — Verification by lattice constant calculation

value of Peak No. 1 to calculate the ratios of the values. The ratios are shown in Column Ratio 1 in Table 5.

When the values of Ratio 1 are multiplied by 3, the obtained values of Ratio 2 become to be simple integers except for Peak Nos. 2 and 3, as can be seen from this table.

According to the detailed chemical analysis, the commercially available salt made from the seawater contains bittern component ($MgCl_2$) as an impurity. Peak Nos. 2 and 3 are reflections from the bittern component. Therefore, we will proceed with the subsequent analysis using the values of Ratio 2 after excluding Peak Nos. 2 and 3. Hereafter, we will employ these simple integer ratios in Ratio 2.

Since the salt crystals look like cubes, we adopt the following "relationship among spacing d_{hkl}, Miller index, and lattice constant" in the following equation:

$$\frac{1}{\mathrm{d}_{hkl}^2} = \frac{h^2+k^2+l^2}{\mathrm{a}^2} \tag{8}$$

Accordingly

$$\frac{1}{\mathrm{d}_{hkl}^2} \infty (h^2+k^2+l^2) \tag{8''}$$

Hence we can index Peak Nos. 1 and 4 ∼ 7 as follows:

Peak No. 1: $(h^2 + k^2 + l^2) = 3 \rightarrow (h\,k\,l) = (1\,1\,1)$

Peak No. 4: $(h^2 + k^2 + l^2) = 4 \rightarrow (h\,k\,l) = (2\,0\,0)$

Peak No. 5: $(h^2 + k^2 + l^2) = 8 \rightarrow (h\,k\,l) = (2\,2\,0)$

Peak No. 6: $(h^2 + k^2 + l^2) = 11 \rightarrow (h\,k\,l) = (3\,1\,1)$

Peak No. 7: $(h^2 + k^2 + l^2) = 12 \rightarrow (h\,k\,l) = (2\,2\,2)$

(Procedure 3) The lattice constant a can be derived from Eq. (8).

$$\therefore\ a = d\sqrt{h^2 + k^2 + l^2} \tag{9}$$

The lattice constants of Peak Nos. 1 and 4 ∼ 7 can be calculated as shown in Table 5. As can be seen from these values, it is apparent that the lattice constant a is constant for each of the peaks within the range of measurement error. Thus, it is confirmed that the indexation in (Procedure 2) mentioned above is correct.

(Procedure 4)

$$\text{Density}\rho = \frac{MZ}{VN} \tag{10}$$

M: Atomic weight (molecular weight)
Z: Number of atoms (number of molecules) in a unit cell
V: Volume of unit cell
N: Avogadro's number

Accordingly,

$$Z = \rho VN/M \tag{10$'$}$$

In this equation, the following values are substituted to obtain the Z value:

Measured density of NaCl $\rho_{\text{obs}} = 2.163\,\text{g/cm}^3$

Volume of unit cell $\text{V} = a^3 = (5.65 \times 10^{-8}\,\text{cm})^3$

$$\text{Avogadro's number} \quad \mathrm{N} = 6.02 \times 10^{23} \text{ pieces/mol}$$

$$\text{Molecular weight} \quad \mathrm{M} = 58.44\,\mathrm{g/mol}$$

$$\therefore\ Z = 4.008 \approx 4.0$$

Therefore, the number of Na and Cl atoms in the unit cell is four each.

(Procedure 5) As already derived at the end of Problem 7, the structural factor of NaCl (fcc) gave the following three results:

(i) When all of h, k, and ℓ are even numbers, $\mathrm{F}(\mathrm{hk}\ell) = 4(f_{Na} + f_{Cl}) \rightarrow$ High intensity
(ii) When all of h, k, and ℓ are odd numbers, $\mathrm{F}(\mathrm{hk}\ell) = 4(f_{Na} - f_{Cl}) \rightarrow$ Low intensity
(iii) When one is odd and the other is even, or when one is even and the other is odd,

$$\mathrm{F}(\mathrm{hk}\ell) = 0 \rightarrow \text{ disappears}$$

Therefore, the extinction rule of fcc can be derived to be ($\mathrm{h}+\mathrm{k} =$ odd number, $k + \ell =$ odd number, $\ell + \mathrm{h} =$ odd number).

We check each of the peak intensities from the above-mentioned relationships between $(\mathrm{hk}\ell)$ and intensity.

Peak No. 1: $(\mathrm{h\,k\,\ell}) = (1\,1\,1)$ corresponds to (ii). Table 4 also shows that the strength is low.
Peak No. 4: $(\mathrm{h\,k\,\ell}) = (2\,0\,0)$ corresponds to (i). Table 4 also shows that the strength is high.
Peak No. 5: $(\mathrm{h\,k\,\ell}) = (2\,2\,0)$ corresponds to (i). Table 4 also shows that the strength is high.
Peak No. 6: $(\mathrm{h\,k\,\ell}) = (3\,1\,1)$ corresponds to (ii). Table 4 also shows that the strength is low.
Peak No. 7: $(\mathrm{h\,k\,\ell}) = (2\,2\,2)$ corresponds to (i). Table 4 also shows that the strength is high.

The difference between Peak No. 1 and No. 7 is subtle, but as mentioned above, it can be roughly divided into those with high strength and those with low strength, which correspond well with the results derived from the structural factors of fcc. In addition, no reflection corresponding to the extinction rule (iii) of fcc was observed.

Therefore, it was confirmed from the structural factors and extinction rules that the NaCl crystal lattice is a face-centered cubic lattice (fcc).

Problem 11. Draw a two-dimensional reciprocal lattice plane using the lattice constant of NaCl, $a = 5.65\,Å$, and show that the (200) and (220) planes indexed in Problem 10, exist just on the lattice points (intersections) in this plane. (cf. Figure 1–25 as a solution example)

A two-dimensional reciprocal lattice plane is drawn as follows. From the lattice constant a $= 5.65\,Å$, the length of one side of the reciprocal lattice can be $a^* = 1/a = 1/5.65 = 0.177 \rightarrow 0.177 \times 10 = 1.77$. Therefore, a two-dimensional tetragonal (square) reciprocal lattice plane with 1.77 cm per scale is drawn as shown in Figure 8. Hereupon, circles having radii of $1/\mathrm{d} \times 10\,(\mathrm{cm})$ are drawn for Peak Nos. 1, 4, 5, 6 and 7 in this reciprocal lattice plane. Then, as can be seen from this figure, we find that Peak Nos. 4 and 5 are on the lattice intersection points of (2 0 0) and (2 2 0) on this plane, respectively.

Problem 12. The crystal of $NiSO_4$ is orthorhombic. The lattice constants are $a = 6.34\,Å$, $b = 7.84\,Å$, $\mathrm{c} = 5.16\,Å$, and the observed density of the crystal was $3.9\,\mathrm{g{\cdot}cm^{-3}}$. Using these values, calculate the number (Z) of molecules included in the unit cell. From this Z value, decide which orthorhombic lattice it has among four lattices of the simple orthorhombic (P), C-centered orthorhombic (C), face-centered orthorhombic (F) and body-centered orthorhombic (I) shown in Figure 1–7.

From the description in this problem, $NiSO_4$ is an orthorhombic crystal, with lattice constants $a = 6.34\,Å$, $b = 7.84\,Å$, $\mathrm{c} = 5.16\,Å$, and the density of this crystal is $3.9\,\mathrm{g/cm^3}$.

From Problem 10 (Procedure 4),

$$Z = \rho \mathrm{VN/M} \qquad (10')$$

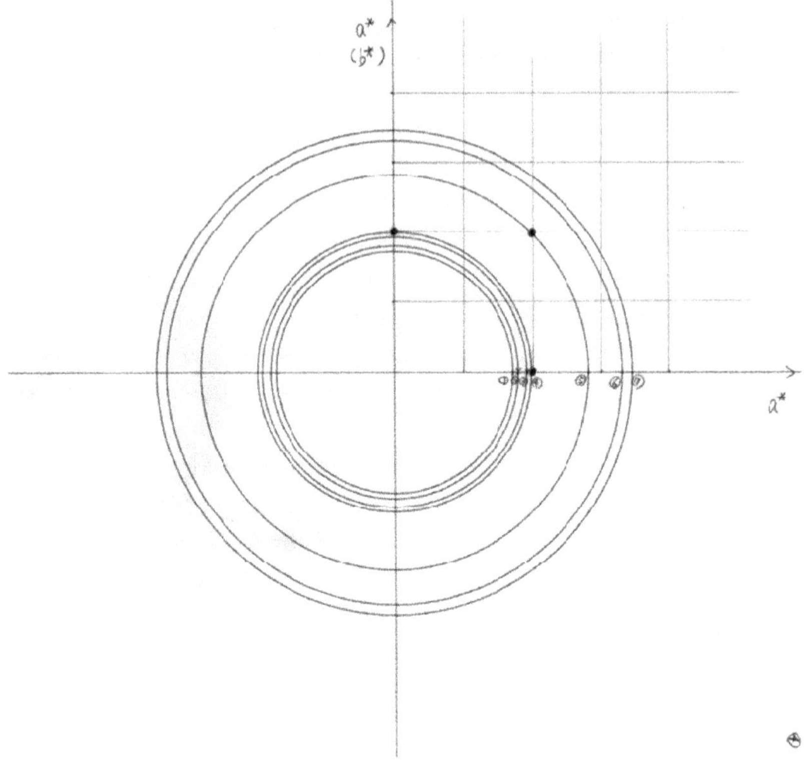

Figure 8. A reciprocal lattice plane of 2D square and the Debye-Scherrer rings of NaCl crystalline powder.

Into this expression we substitute the following values:

Measured density of $NiSO_4$ $\rho_{\text{obs}} = 3.9\,\text{g/cm}^3$

Unit cell volume $\text{V} = abc = (6.34 \times 10^{-8}\,\text{cm}) \times (7.84 \times 10^{-8}\,\text{cm}) \times (5.16 \times 10^{-8}\,\text{cm}) = 256.5 \times 10^{-24}\,\text{cm}^3$

Avogadro's number $\text{N} = 6.02 \times 10^{23}\,\text{pieces/mol}$

Molecular weight $\text{M} = 155.00\,\text{g/mol}$

$$\begin{aligned}\therefore\ Z &= 3.9(\text{g/cm}^3) \times 256.5 \times 10^{-24}(\text{cm}^3)\\ &\quad \times 6.02 \times 10^{23}(\text{pieces/mol}))/(155.00\,(\text{g/mol}))\\ &= 3.9\\ &\cong 4\ (\text{pieces})\end{aligned}$$

When the case of cubic crystals, the Z values should be

Simple cubic (P): Z = 1

C-base-centered cubic (C): Z = 2

Face-centered cubic (F): Z = 4

Body-centered cubic (I): $Z = 2$

Therefore, the cubic system of $NiSO_4$ is a face-centered cubic system (F) because Z = 4 as calculated above.

Problem 13. Calculate the filling factors for the following unit cells. (a) Simple cubic lattice, (b) body-centered cubic lattice (bcc), (c) face-centered cubic lattice (fcc).

(a) Simple cubic lattice
The unit cell of a simple cubic lattice is shown in Figure 9(a).

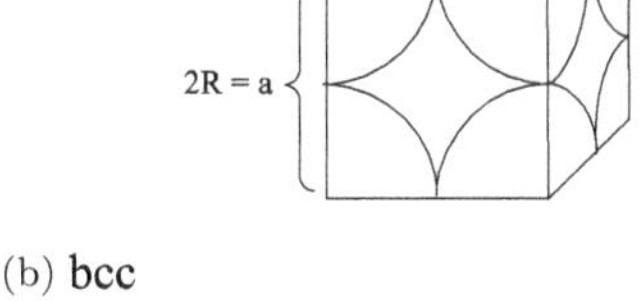

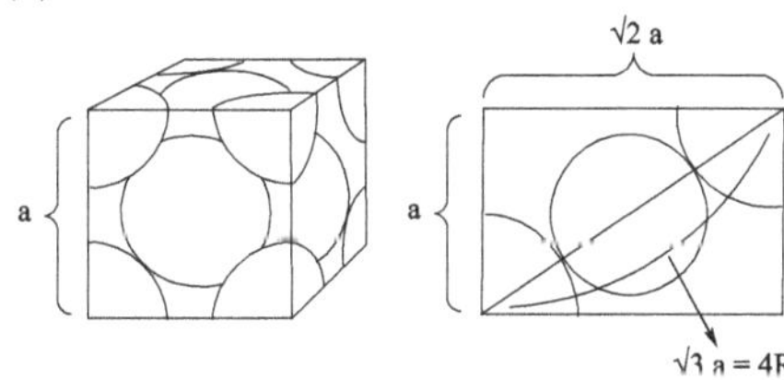

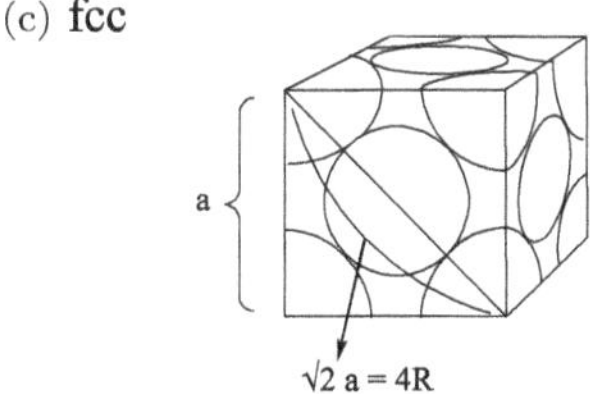

Figure 9. Unit cells: (a) simple cubic lattice, (b) body-centered cubic lattice, and (c) face-centered cubic lattice. R = radius of atom, a = lattice constant of cubic lattice.

Therefore,
(Filling factor) = (atomic volume)/(unit cell volume)

$$= \frac{\left(\frac{4}{3}\pi R^3 \times \frac{1}{8}\right) \times 8}{a^3}$$

$$= \frac{\frac{4}{3}\pi \left(\frac{1}{2}a\right)^3}{a^3} \quad (\because 2R = a)$$

$$= \frac{\pi}{6}$$

$$\cong 0.5236$$

(b) Body-centered cubic lattice
The unit cell of a body-centered cubic lattice is shown in Figure 9(b). Therefore,

$$(\text{Filling factor}) = \frac{\frac{4}{3}\pi R^3 \left(1 + \frac{1}{8} \times 8\right)}{a^3}$$

$$= \frac{\frac{4}{3}\pi \left(\frac{\sqrt{3}}{4}a\right)^3 \times 2}{a^3} \quad (\because 4R = \sqrt{3}a)$$

$$= \frac{\sqrt{3}}{8}\pi$$

$$\cong 0.6802$$

(c) Face-centered cubic lattice
The unit cell of a face-centered cubic lattice is shown in Figure 9(c). Therefore,

$$(\text{Filling factor}) = \frac{\frac{4}{3}\pi R^3 \left(\frac{1}{2} \times 6 + \frac{1}{8} \times 8\right)}{a^3}$$

$$= \frac{\frac{4}{3}\pi \left(\frac{\sqrt{2}}{4}a\right)^3 \times 4}{a^3} \quad (\because 4R = \sqrt{2}a)$$

$$= \frac{\sqrt{2}}{6}\pi$$

$$\cong 0.7405$$

Chapter 2

Solutions to End-of-Chapter Problems

Problem 1. Explain briefly what is polymorphism. Also give examples of polymorphisms.

Polymorphism is the phenomenon in which any substance or compound has more than one crystalline form. Different crystalline forms of the same compound have different physical properties such as density, hardness, optical and electrical properties, solubility, and drug efficacy. Polymorphs are also crystalline phases of a compound that result from at least two different possible molecular arrangements in the crystalline state. The molecular structure is unchanged in the two polymorphs.

Examples of polymorphs are given below.

- Carbon (C_∞)
  Diamond: cubic system, hardness 10, generally an insulator at room temperature, high resistance ($10^{13}\,\Omega\,\mathrm{cm}$)
  Graphite: hexagonal system, hardness 1-2, semi-metallic, low resistance (about $10^{-3}\,\Omega\,\mathrm{cm}$)
  See Figure 1
- Ice (H_2O)
  Figure 2 shows the phase diagram of H_2O. As you can see, ice has many polymorphs. Table 1 summarizes these polymorphic crystalline forms of ice. As can be seen from this table, it should be noted that although different polymorphs exhibit the same crystal system, they each have different space groups.

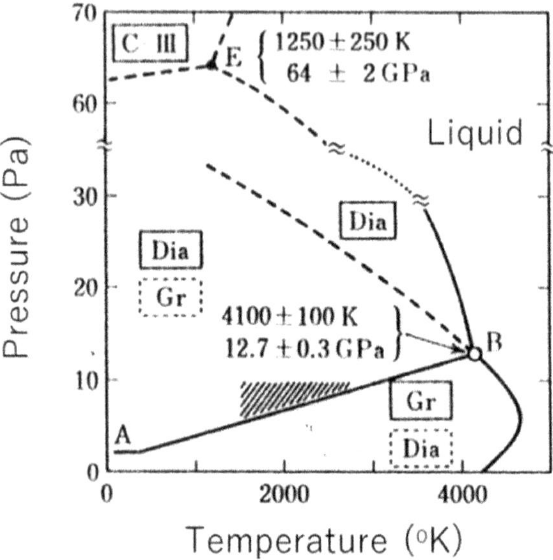

Figure 1. Phase diagram of carbon.

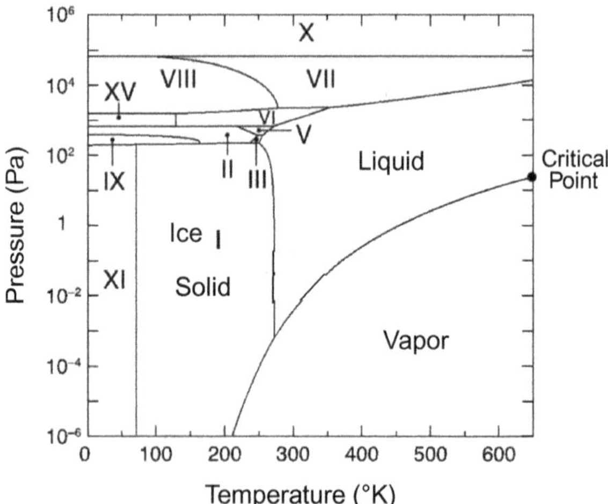

Figure 2. Phase diagram of H_2O.

Table 1. Polymorphs of H_2O.

Polymorph	Crystal system
I	Cubic
II	Rhombohedral
III	Tetragonal
IV	Rhombohedral (metastable)
V	Monoclinic
VI	Tetragonal
VII	Cubic
VIII	Tetragonal
IX	Tetragonal
X	Cubic
XI	Orthorhombic
XII	Tetragonal (metastable)
XIII	Monoclinic
XIV	Orthorhombic
XV	Pseudo- orthorhombic
XVI	Cubic
XVII	Hexagonal

- Sulfur (crown molecule S_8)
  Orthorhombic S_α and monoclinic S_β
  Figure 3 shows the phase diagram of sulfur.
- ZnS
  Wurtzite (Hexagonal) and Sphalerite (Cubic).
- Medicine
  Sometimes they give more than 10 polymorphs. Each polymorph has different solubility, some with high drug efficacy, some with low drug efficacy, some with fast-acting effect, and some with slow-acting effect. Therefore, polymorphism is an important issue for the pharmaceutical science and technology. (See Problem 18)
- Soap
- Paraffin

(Remarks) Polymorphs can be seen as the different crystalline forms in the crystalline state (= solid state), but liquid and gas states do not show polymorphism: they are identical. (Very rare exceptions: liquid polymorphs can be seen only in liquid helium I and liquid helium II; low density water and high density water)

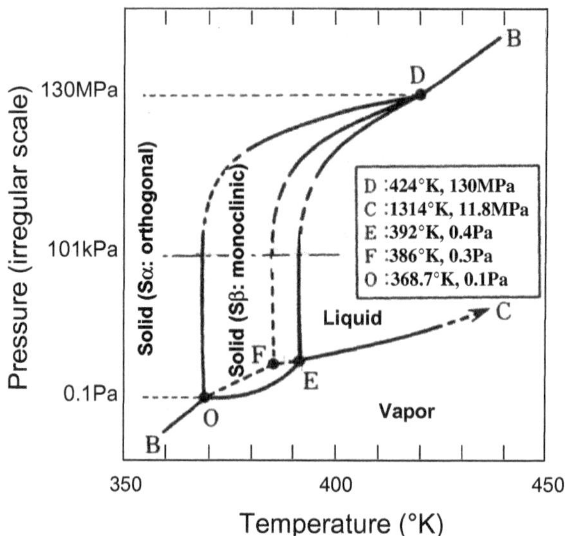

Figure 3. Phase diagram of sulfur.

Problem 2. How can we distinguish between polymorphs and tautomers?

Polymorphs and tautomers can be distinguished by melting the material. As remarked at the end of the answer for Problem 1, polymorphs can be seen as the different crystalline forms in the crystalline state, but liquid and gas states do not show polymorphism: they are identical. Therefore, the solid polymorphs give different molecular assemblies in the solid state, but their melts give the same liquid state having the identical properties. On the other hand, tautomers give different own liquid phases with different properties, *e.g.*, refractive index. This difference allows us to distinguish between polymorphs and tautomers.

Problem 3. Describe the features and differences of the first-order phase transition and the second-order phase transition.

- Characteristics of first-order phase transition

The first derivative of the chemical potential is discontinuous.

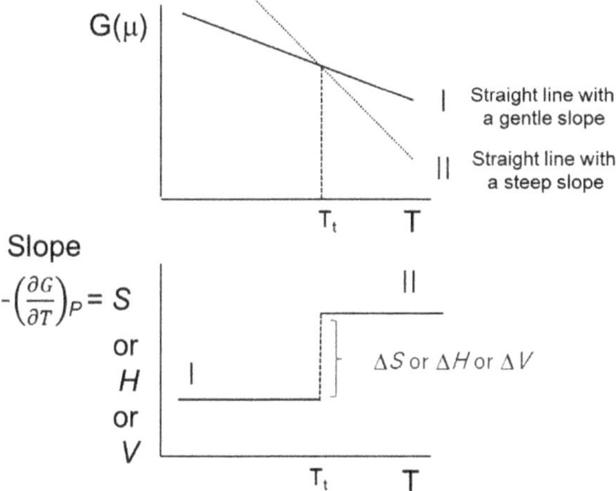

Figure 4. Changes in thermodynamic properties associated with the first-order phase transition.

As shown in Figure 4, each polymorph is represented by a straight line in the G–T diagram. The first-order phase transition occurs at the intersection of two straight lines with different slopes. In this first-order phase transition, each of the S–T, H–T and V–T diagrams shows that S, H, and V are discontinuous at the phase transition temperature T_t. From these diagrams, at the phase transition temperature T_t we can see the following three characteristics:

(1) the entropy S suddenly changes (into a disordered state or an ordered state),
(2) the enthalpy H suddenly changes the heat content (bigger or smaller),
(3) the volume V suddenly changes (expands or shrinks).

- Characteristics of the second-order phase transition.
  The second derivative of the chemical potential is discontinuous.
  As shown in the G–T diagram in Figure 5, each polymorph is represented by a curve. The second-order phase transition occurs at the intersection of two curves with different curvatures. Also, as can be seen from the slope(S) –T diagram, each slope of the two curves is exactly the same at this phase transition temperature T_t.

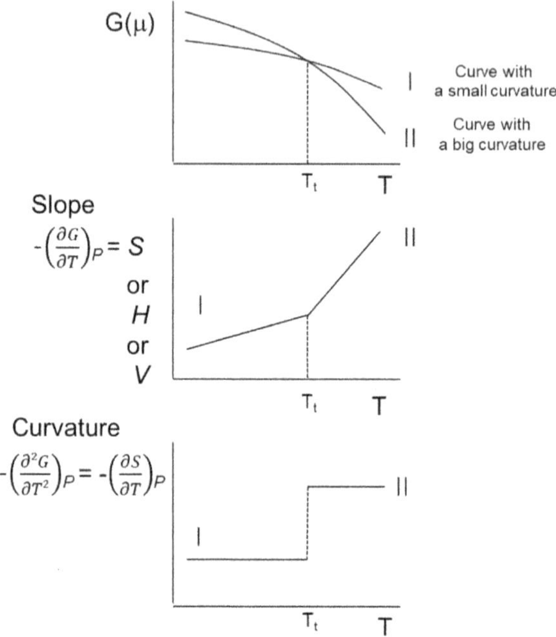

Figure 5. Changes in thermodynamic properties associated with the second-order phase transition.

Furthermore, as can be seen from the curvature-T diagram, the second derivative of the chemical potential is discontinuous at the phase transition temperature T_t.

There are two main causes of the second-order phase transitions: order-disorder transitions and rotational transitions.

Problem 4. Give examples of the second-order phase transition.

Examples of the second-order phase transitions are listed as follows:

(a) Rotational transition of molecular crystals
(b) Order-disorder transition of alloys
(c) Phase transition of ferroelectrics
(d) Magnetic phase transitions such as ferromagnetism and anti-ferromagnetism

(e) λ-transition of liquid helium
(f) Phase transition of superconductor

and so on.

A common feature of second-order phase transitions is that, although there is no change in the crystal structure at the transition temperature, anomalous λ-type specific heat is observed, and a certain order, which does not exist at higher temperatures, appears in the low-temperature phase. Therefore, we can broadly categorize the causes of second-order phase transition into two cases: "order-disorder transition" [examples from (b) to (f) listed above] and "rotational transition" [example (a) listed above].

Concrete examples of "order-disorder transition" and "rotational transition" are shown below.

Examples of "order-disorder transition"

- Phase transition of KCN crystal

KCN is an ionic crystal and can be expressed as $K^+(C \equiv N)^-$. The CN anion has two states: $C \equiv N$-oriented and $N \equiv C$-oriented.

At lower temperatures, all number (N) of anions are $C \equiv N$-oriented or $N \equiv C$-oriented.

Therefore, since there is only one possible state, the entropy S_ℓ at this time is given by $S = k\ell nW$ (W: number of states)

$$S_l = k\ell n1 = 0$$

On the other hand, at higher temperatures, $C \equiv N$-oriented : $N \equiv C$-oriented $= 1 : e^{-\frac{\Delta E}{kT}}$

Above the phase transition temperature, the numbers for both orientations are equal. Therefore, since there are 2^N states, the entropy S_h at this time is

$$S_h = k\ell n2^N = Nk\ell n2 = R\ell n2$$
$$= 1.38\,\text{cal}/(\text{deg} \cdot \text{mol})$$

$$\therefore \Delta S = S_h - S_l = 1.38\,\text{cal}/(\text{deg} \cdot \text{mol}) \text{.................(calculated value)}$$
$$1.32\,\text{cal}/(\text{deg} \cdot \text{mol}) \text{.................(measured value)}$$

Therefore, the calculated value and the measured value agree within the experimental error, proving that this phase transition is an "order-disorder transition."

- Examples of "rotational transitions"

The rotational transition is observed for the following bulky and spherical molecules such as methane (CH_4), silane (SiH_4), carbon tetrachloride (CCl_4), and fullerenes (C_{60}). The phase transition is known as "crystal-plastic crystal phase transition" which is the second-order phase transition.

Problem 5. Explain monotropic and enantiotropic relationship on P–T diagram and G–T diagram.

Figure 6 shows a G–T diagram [A] and a P–T diagram [B] showing the enantiotropic relationship between polymorphs I and II.

(5-1) G–T diagram for enantiotropic relationship between I and II

First, we look at the G–T diagram in Figure 6A. On the lower temperature region in this G–T diagram, the line of Polymorph I is lower than the line of Polymorph II, so that Polymorph I has a lower chemical potential (= Gibbs free energy G) than Polymorph II. Therefore, Polymorph I is more stable than Polymorph II at these lower temperatures. When you stand on this crystal polymorph I line as illustrated in the figure. As this stable polymorph I is slowly heated up, *i.e.*, as you walk to the right on Line I in the G–T diagram, you encounter the intersection of Line I and Line II at T_t. Crystalline Polymorph II is more stable than Polymorph I on the right side of this intersection, that is, on the higher temperature region. Accordingly, you completely transform into Polymorph II at this intersection of T_t. On further heating, you move to the right on Line II, and then, you encounter the intersection of Line II and Line I.L. at T_{m2}. In the right side of this intersection, *i.e.*, on the higher temperature region, Isotropic liquid I.L. is more stable than Polymorph II. Accordingly, your Polymorph II completely transforms into the isotropic liquid I.L.

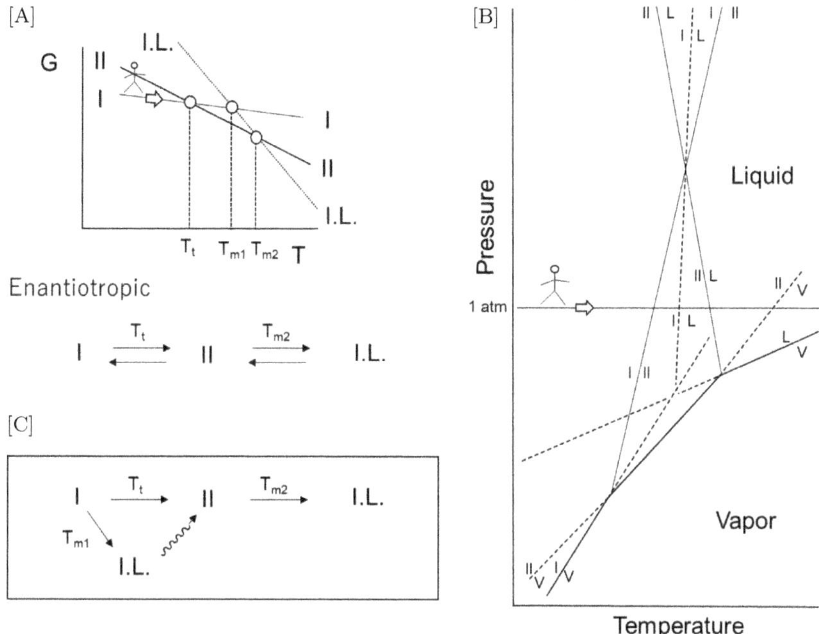

Figure 6. G–T diagram [A] and P–T diagram [B] for monotropic relationship between crystalline polymorphs I and II.

Thus, in a slow heating process, the phase transitions are observed like as I → II → I.L.

On the other hand, when this isotropic liquid I.L. is slowly cooled, you walk to the left on the straight line I.L., *i.e.*, to the lower temperature side in Figure 6A, and then you encounter the intersection of Line I.L. and Line II at T_{m2}. In the left part of this intersection, *i.e.*, on the lower temperature region from T_{m2}, Polymorph II is more stable than isotropic liquid I.L. Therefore, you transform from I.L. to Polymorph II at this intersection of T_{m2}. On further cooling, you walk to the left on straight line II and encounter the intersection of Line II and Line I at T_t. In the left side of this intersection, *i.e.*, on the lower temperature region, Polymorph I is more stable than Polymorph II. Therefore, you completely transform to Polymorph I at this intersection at T_t, and then walk to the left on Line I at these lower temperatures. Thus, in this slow cooling process, the phase transitions are observed like as I.L. → II → I.

Therefore, in the entire heating and cooling process, the phase transition occurs as I⇄II⇄I.L. In this way, the phase transition has a mirror image relationship between the heating process and the cooling process. Hence, polymorphs I and II are termed to have an enantiotropic relationship in the sense that they are thermally mirror images.

However, the above is the case when the heating rate is slow enough. If the heating rate is very fast, another phenomenon occurs. It is the double melting behavior due to superheating [Figure 6C].

When Polymorph I stable in the lower temperature region is **rapidly** heated, you of Polymorph I run fast to the right on Line I in the G–T diagram[A] and encounter the intersection of Line I and Line II at T_t. However, the heating rate is so fast that you cannot completely transform to Polymorph II at this intersection of T_t, and the remaining Polymorph I is superheated to run on Line I, as it is, to reach to the intersection of Line I and Line I.L. at T_{m1}. Hereupon, you remaining as crystalline Polymorph I transform into isotropic liquid I.L. Since this I.L. is less stable than crystalline Polymorph II, you of the I.L. **rapidly** transform to Polymorph II by using the seed crystals formed partially at the intersection at T_t. Thus, we can observe the melting point of Polymorph I (T_{m1}) and the rapid relaxion from this isotropic liquid I.L. to crystalline Polymorph II. When the temperature is further elevated, you of Polymorph II encounter the intersection of Line II and Line I.L. at the melting point of Polymorph II (T_{m2}), where you again transform to the I.L. Therefore, when heated **rapidly**, two times of melting behavior can be observed on one-heating run [Figure 6C].

(5-2) G–T diagram for monotropic relationship between I and II

Figure 7 illustrates [A] a G–T diagram and [B] a P–T diagram when polymorphs I and II have a monotropic relationship.

First, we look at the G–T diagram in Figure 7A. In the temperature region lower than the virtual intersection at T_t, Line I is lower than Line II. Therefore, crystalline Polymorph I has a lower chemical

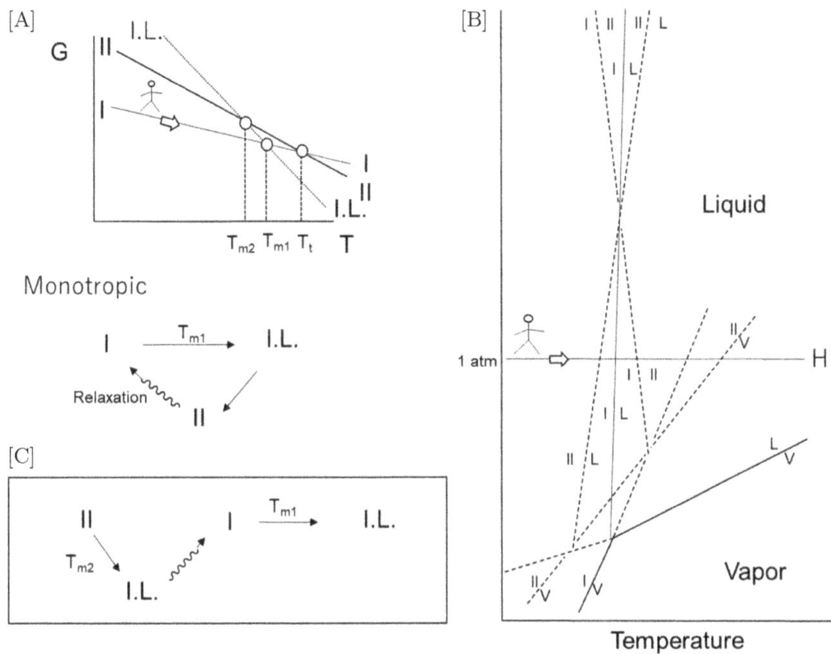

Figure 7. G–T diagram [A] and P–T diagram [B] for enantiotropic relationship between crystalline polymorphs I and II.

potential (= Gibbs free energy G) than crystalline Polymorph II, *i.e*, Polymorph I is more stable than Polymorph II at these temperatures. When you stand on this stable polymorph I, as illustrated in this figure. As this stable polymorph I is heated, you walk to the right on Line I and then encounter the intersection of Line I and Line I.L. at T_{m1}. At this point Polymorph I completely melts into isotropic liquid I.L. Thus, in this heating process, the phase transition occurs like as I $\rightarrow$ I.L.

On the other hand, when this isotropic liquid I.L. is rapidly cooled, you of I.L. run to the left on Line I.L. without transforming to Polymorph I at T_{m1}, *i.e.*, the I.L. is supercooled and reaches to the intersection of Line I.L. and Line 2 at T_{m2}. Hereupon, you of I.L. transform into Polymorph II at this intersection of T_{m2}; on further cooling, you run to the left on Line II. Crystalline Polymorph II is less stable than Polymorph I in the lower temperature region below this intersection at T_{m2}, so that you of Polymorph II slowly relax to more stable Polymorph I. After relaxing, you of Polymorph I run to the left on Line I. When rapidly cooled, the phase transition sequence is, thus, observed like as I.L. $\rightarrow$ II $\sim\sim\rightarrow$ I.

Therefore, the cooling process does not take the same path as the heating process as illustrated in Figure 7A. Accordingly, Polymorphs I and II are termed to have a monotropic relationship in the sense that they are in a one-way thermal relationship.

However, the above is when the stable polymorph I in a monotropic relationship between I and II is heated from the lower temperature. When the unstable polymorph II is heated from the lower temperature, another phenomenon occurs. It is a double melting behavior accompanied by relaxation depicted in Figure 7C. As illustrated in the G–T diagram [Figure 7A], Line I is lower than Line II in the temperature region lower than the virtual intersection at T_t, so that the crystalline polymorph I has a lower chemical potential (= Gibbs free energy G) than the crystalline polymorph II, *i.e.*, Polymorph I is more stable than Polymorph II at room temperature. Nevertheless, the stable polymorph I is not the only one that exists in the real world at room temperature. It is remarkable that such an unstable polymorph II can be often obtained at room temperature depending on the conditions of recrystallization (see Q14 for details).

When the obtained unstable polymorph II is **slowly** heated from room temperature, you of Polymorph II walk to the right on Line II in the G–T diagram [Fig. 7A] and reach to the intersection of Line II and Line I.L. at T_{m2}. Hereupon the crystalline polymorph II completely melts into the isotropic liquid I.L. However, at this temperature the isotropic liquid I.L. is less stable than the crystalline polymorph I. Therefore, you can't stay as the isotropic liquid I.L., and then you of I.L. **slowly** transform to the crystalline polymorph I by relaxation, *i.e.*, recrystallization occurs. This recrystallization is very slow, because there is no real solid-solid phase transition in the temperature region lower than T_{m1} and T_{m2}, and because the seed crystals of Polymorph I do not co-exist in this I.L., contrary to the case of the enantiotropic relationship between I and II mentioned above. If Polymorph II is rapidly heated, the recrystallization may hardly occur. After enough time to complete the recrystallization, you of Polymorph I walk to the right on Line I to reach the intersection of Line I and Line I.L. at T_{m1}. Hereupon, you transform again to the I.L., *i.e.*, you melt again. Thus, therefore, when heated **slowly**, two times of melting behavior can be observed on one-heating run [Figure 7C]. If the heating rate is fast enough, no relaxation to polymorph I occurs and no double melting behavior is observed.

From Sections (1) and (2) mentioned above, we can find the following interesting facts. When Polymorphs I and II are in an **enantiotropic relationship**, the double melting behavior can be observed **only for the rapid heating** [Figure 6C]. On the other hand, when Polymorphs I and II are in **monotropic relationship**, the double melting behavior can be observed **only for slow heating** [Figure 7C].

(5-3) P–T diagram for enantiotropic relationship between I and II

Next, we look at a P–T diagram illustrated in Figure 6B. In this figure, solid and dotted lines indicate phase boundaries: the solid lines and the dotted lines represent actual phase transitions and virtual phase transitions, respectively. Then we draw a horizontal line at ambient pressure (1 atm) in this figure and you stand at a lower temperature on this line, so that you are crystalline Polymorphic I. When you walk slowly from the left to the right on the horizontal line, *i.e.*, when Polymorph I is slowly heated under 1 atm, you first encounter the boundary of Polymorph I and Polymorph II, *i.e.*, a solid-solid phase transition point. Hereupon, you transform from the crystalline Polymorph I to the crystalline Polymorph II. When you of Polymorph II walk further to the right, you reach to the borderline of Polymorph II and liquid phase L. Hereupon, you of Polymorph II transform to the liquid phase L, *i.e.*, the melting of Polymorph II occurs. When you of the liquid phase L turn back to walk slowly to the left, *i.e.*, when the liquid L is slowly cooled under 1 atm, you encounter the boundary of liquid L and crystalline Polymorph II. Hereupon, you of L transform here to crystalline Polymorph II. If you walk further to the left, you come across the borderline of Polymorph II and Polymorph I, *i.e.*, the solid-solid phase transition point. Here, you of crystalline Polymorph II transform back to the original crystalline Polymorph I.

Thus, when Polymorph I and Polymorph II are in an enantiotropic relationship, we can realize also from the P–T diagram [Figure 6B] that the phase transition occurs like as I⇄II⇄I.L, as could be previously seen in the corresponding G–T diagram [Figure 6A].

However, the above is the case when the heating rate is slow enough. When the heating rate is fast enough, we can observe another phenomenon of the double melting behavior due to superheating: it can be explained even from the P–T diagram in Figure 6B, as it has already seen from the G–T diagram in Figure 6A.

In Figure 6B, when the stable crystalline polymorph I is rapidly heated from a low temperature under 1 atm, you of Polymorph I run to the right along the horizontal line in this figure to reach the boundary of Polymorph I and Polymorph II, *i.e.*, the solid-solid phase transition point. However, you of Polymorph I cannot completely transform to Polymorph II due to the fast heating. On further heating at the fast rate, the remained Polymorph I runs to the right on the horizontal line, as it is, and then reach to the boundary of crystalline Polymorph I and Liquid L denoted by a dotted line in this figure. Here, the remained crystalline Polymorph I melts to the liquid L. Since this liquid L is less stable than Polymorph II, the liquid L rapidly recrystallizes into the crystalline Polymorph II, by using the seed crystals of Polymorph II partially formed at the precedent solid–solid phase transition, and then you completely transform to the crystalline Polymorph II. When the temperature is further increased, you of Polymorph II encounter the transition boundary of Polymorph II and liquid L. Hereupon, you of Polymorph II transform into liquid phase L again. Thus, the double melting behavior can be realized also from the P–T diagram [Figure 6B].

(5-4) P–T diagram for monotropic relationship between I and II

Next, we look at the P–T diagram illustrated in Figure 7B. In this figure, solid and dotted lines indicate phase boundaries: the solid lines and the dotted lines represent actual phase transitions and virtual phase transitions, respectively. Then we draw a horizontal line at ambient pressure (1 atm) in this figure and you stand at a lower temperature on this line, so that you are crystalline Polymorphic I. When you walk from the left to the right on the horizontal line, *i.e.*, when Polymorph I is heated under 1 atm, you first encounter the boundary of Polymorph I and liquid L, *i.e.*, the melting point of crystalline Polymorph I. Hereupon, you of Polymorph I transform to the liquid L. When you of liquid L turn back to walk to the left, *i.e.*, when it is cooled under 1 atm, you reach to the boundary of the liquid L and the crystalline Polymorph I. Hereupon, you of liquid L transform to the original Polymorph I. Thus, the phase transition occurs like as, **I⇄I.L**. on a heating and cooling cycle for the monotropic relationship between I and II.

However, the above is when the stable crystalline Polymorph I is heated. When the unstable crystalline Polymorph II is heated from a low temperature, another phenomenon of double melting behavior occurs, as illustrated in Figure 7C. We can realize the double melting behavior also from this P–T diagram [Figure 7B], as could be previously seen in the corresponding G–T diagram [Figure 7A].

When you of Polymorph II slowly walk to the right from a low temperature on the horizontal line in the P–T diagram [Figure 7B], you encounter the phase boundary (denoted by a dotted line) of Polymorph II and the liquid L. Hereupon, Polymorph II completely melts to the liquid L. However, since the liquid L at this temperature is less stable than the crystalline Polymorph I, the liquid L slowly recrystallizes to the crystalline polymorph I. This slow recrystallization is due to lack of the seed-crystals Polymorph I in the liquid L. It is resulted from no solid–solid phase transition below the melting points of Polymorph I and II. When you of the completely recrystallized polymorph I walks to the right on the horizontal line, you encounter the boundary (solid line) of Polymorph I and liquid L phase boundary. Hereupon, you melt again to the liquid L. Thus, when the unstable crystalline Polymorph II is slowly heated from a low temperature, the double melting behavior could be explained also from the P–T diagram [Figure 7B]. If the heating rate is fast enough, no double melting behavior is observed, because it has no time enough to recrystallize from liquid L to Polymorph I.

On the other hand, when the liquid L is rapidly cooled from a high temperature under 1 atm, you of the liquid L run fast to the left along the horizontal line in Figure 7B and then encounter the boundary of liquid L and crystalline Polymorph I. However, the liquid L does not recrystallize into Polymorph I, because the heating rate is so fast that it has not enough to recrystallize. On further running to the left along the horizontal line, you reach to the boundary of liquid L and crystalline Polymorph II. Hereupon, you of the liquid L may partially recrystallize into the unstable Polymorph II. If the cooling rate is so fast, the liquid L does not recrystallize into Polymorph II. Such a supercooling results in an unstable polymorph II at room temperature, suppressing the relaxation from the unstable polymorph II to the stable polymorph I. This is because rearrangement of the molecules is difficult in solid at low temperatures. Therefore, it can be seen that rapid cooling facilitates the formation of unstable polymorph II.

Problem 6. Discuss relaxation.

Relaxation is the fall from a metastable phase to a stable phase corresponding a wavy line in G–T diagrams illustrated in Figure 8. However, relaxation is not a phase transition: relaxation occurs at an indefinite temperature (T_x), whereas each of the phase transitions occur at a certain definite temperature under atmospheric pressure, *e.g.*, T_{m1} and T_{m2} in Figure 8. When T_{m1} and T_{m2} are very close, whether heat is absorbed or exothermed during relaxation depends on the slope difference ($\propto \Delta S = S_2 - S_1$) between the crystalline polymorphs K_1 and K_2 lines in the G–T diagram.

The problem of exothermic or endothermic relaxation can be proved as follows. Figure 8 schematically shows the slopes -S_1 and

-S_2 of polymorphs 1 (K_1) and 2 (K_2) when [A] $S_1 < S_2$ and [B] $S_1 > S_2$. We will consider these cases.

(Proof)

$$G = H - TS$$
$$\therefore S = \frac{H - G}{T}$$

When relaxation occurs at a temperature Tx,

$$S_1 = \frac{H_1 - G_1}{T_X}, \quad S_2 = \frac{H_2 - G_2}{T_X}$$
$$\therefore \Delta S = S_2 - S_1 = \frac{H_2 - G_2}{T_X} - \frac{H_1 - G_1}{T_X}$$
$$= \frac{(H_2 - H_1) - (G_2 - G_1)}{T_X}$$
$$= \frac{\Delta H - \Delta G}{T_X}$$

For Case [A],
$S_1 < S_2$
$\Delta S = S_2 - S_1 > 0$
$\therefore \Delta H - \Delta G > 0$
$\therefore \Delta H > \Delta G$

For Case [B],
$S_1 > S_2$
$\Delta S = S_2 - S_1 < 0$
$\therefore \Delta H - \Delta G < 0$
$\therefore \Delta H < \Delta G$

Here, relaxation from Line K_1 to Line K_2 tends to occur when Lines K_1 and K_2 are very close to each other, *i.e.*, when T_{m1} and T_{m2} are very close. Therefore, in such a case $\Delta G \approx 0$ can be considered.

For Case [A],
$\therefore \Delta H > 0$
$\therefore$ Endothermic relaxation occurs.
(rarely observed)

For Case [B],
$\therefore \Delta H < 0$
$\therefore$ Exothermic relaxation occurs.
(most of the cases)

(End of proof)

Thus, it is very interesting that just by looking at the difference between the slopes of the two polymorphic straight lines ($\propto \Delta S$) in

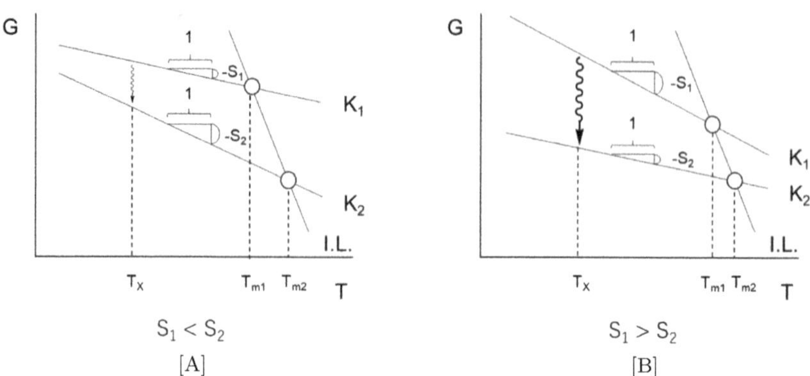

Figure 8. G–T diagrams for [A] endothermic relaxation and [B] exothermic relaxation.

the G–T diagram, we can intuitively know whether it absorbs heat or generates heat when the relaxation (Figure 8). We observe mostly exothermic relaxation [Case B], but rarely endothermic relaxation [Case A].

Problem 7. Draw the HMX G–T diagram (Figure 2.15 in the textbook) and explain the phase transitions of each polymorph in the heating process. Also, explain how each polymorph forms when supercooling occurs during the cooling process.

If only the most stable states exist in a heating process accompanied by the multiple phase transitions, the phase transition must be as follows:

$$\text{115.5°C 165.5°C } \mathrm{T_{m4}}$$

$$\mathrm{I}(\alpha) \rightarrow \mathrm{II}(\beta) \rightarrow \mathrm{IV}(\delta) \rightarrow \mathrm{I.L.}$$

When supercooling occurs in the cooling process, each crystalline polymorphs precipitates depending on the degree of supercooling, as follows:

- In the case that Polymorph I (α) precipitates: when the I.L. higher than $\mathrm{T_{m4}}$ is slowly cooled, it recrystallizes to Polymorph IV (δ) at $\mathrm{T_{m4}}$; on further slow cooling, it transforms to Polymorph II (β). When it is cooled further slowly to below 115.5°C, it remains in polymorph II (β) in a supercooled state for a while, but relaxes to Polymorph I (α) at a certain temperature. This relaxation temperature is indefinite.

- In the case that Polymorph II (β) precipitates: when the I.L. higher than T_{m4} is slowly cooled, it recrystallizes to polymorph IV (δ) at T_{m4}; on further slow cooling and then the temperature is kept between 165.5°C > T > 157°C, the supercooled Polymorph IV (δ) relaxes to Polymorph II (β). When it is rapidly cooled to room temperature, *i.e.*, quenched it to room temperature, Polymorph II (β) can be obtained without relaxation.
- In the case that Polymorph III (γ) precipitates: when the I.L. higher than T_{m4} is slowly cooled, it recrystallizes to Polymorph IV (δ) at T_{m4}. When Polymorph IV (δ) is supercooled slightly below 132°C and kept at this temperature, it recrystallizes to Polymorph III (γ) by relaxation. Quenching it to room temperature gives Polymorph III (γ) without further relaxation.
- In the case that Polymorph IV (δ) precipitates: when the I.L. higher than T_{m4} is kept between $T_{m4} > T > T_{m3}$, the supercooled I.L. recrystallizes to Polymorph IV (δ) by relaxation. Quenching it to room temperature gives Polymorph IV (δ) without further relaxation.

Thus, each polymorph can be obtained by the different supercooling method.

Problem 8.

(i) Explain the solubility curve and the solution phase transition.
(ii) Why crystal polymorph having lower melting shows higher solubility?
(iii) Explain thermodynamically why the solubility curve intersection corresponds to the solid–solid transition temperature.

(i)

Solubility curve

It is a curve of the solubility versus temperature, which represents the limited masses of solute dissolved in a solvent depending on the temperatures. We often use one of the following three definitions of solubility: (i) the mass (g) of solute per 100 g of solvent; (ii) the amount (g) of solute in 100 g of solution; (iii) the mass (g) of solute dissolved in 1L (= dm^3 = 1000 ml) of solution.

Solution phase transition

When two different crystalline polymorphs of the same compound are placed in a saturated solution, the more soluble polymorph dissolves and, on the contrary, the less soluble polymorph grows. This phenomenon is called solution phase transition. We depict the stabilities of two crystalline polymorphs I and II in Figure 9 by using a G–T diagram. In this case, the appearance of the solution phase transition is as shown in the circles. The unstable polymorph has higher solubility: (i) when $G_I < G_{II}$, *i.e.*, in the temperature region lower than 115.5°C, Polymorph II has higher solubility so

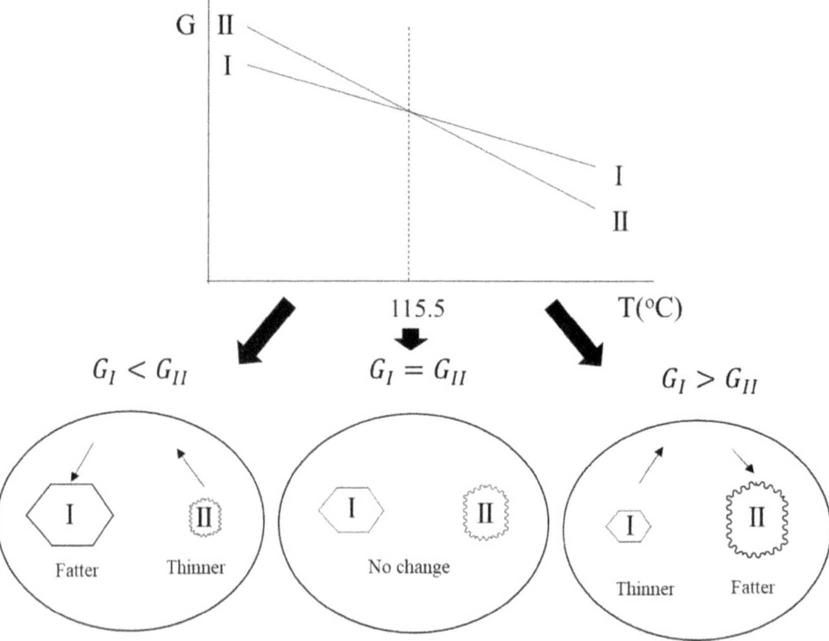

Figure 9. An example of solubility curves of crystalline polymorphs I and II. The solubilities of the crystalline polymorphs in a solution depends on the stabilities in the corresponding G–T diagram.

that Polymorph II dissolves but Polymorph I grows; (ii) when $G_I > G_{II}$, *i.e.*, in the temperature region higher than 115.5°C, Polymorph I has higher solubility so that Polymorph I dissolves but Polymorph II grows; (iii) when $G_I = G_{II}$, *i.e.*, at the solid–solid phase transition temperature between the two polymorphs, their solubility is equal and we cannot observe any change in the shape of these two polymorphs I and II. Thus, a solid–solid phase transition can be detected also by the appearance change in a solution.

(ii)

In general, the G–T diagram of a compound having multiple crystalline polymorphs can be drawn, *e.g.*, as shown in Figure 10[A]. In this example, Polymorph I is the most stable and Polymorph IV is the least stable. The lower the stability becomes in the G–T diagram, the less stable the polymorph has, *i.e.*, the lower the melting point becomes like as $T_{m4} > T_{m3} > T_{m2} > T_{m1}$.

Therefore, for the compound showing a G–T diagram like Figure 10[A], the solubility (μ)-T diagram can be drawn as Figure 10[B]. As can be seen by comparing Figures 10[A] and [B], if the order of the lines from I to IV in Figure 10[A] are swapped top and bottom, and if the downward-sloping straight lines are changed into the curves with upward-curvatures, Figure 10[B] can be obtained.

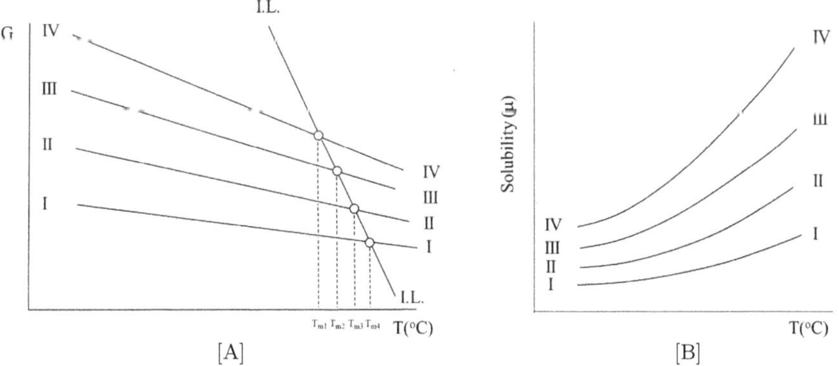

Figure 10. Schematic representations of a G–T diagram [A] and the corresponding solubility(m)–T diagram [B] for four crystalline polymorphs.

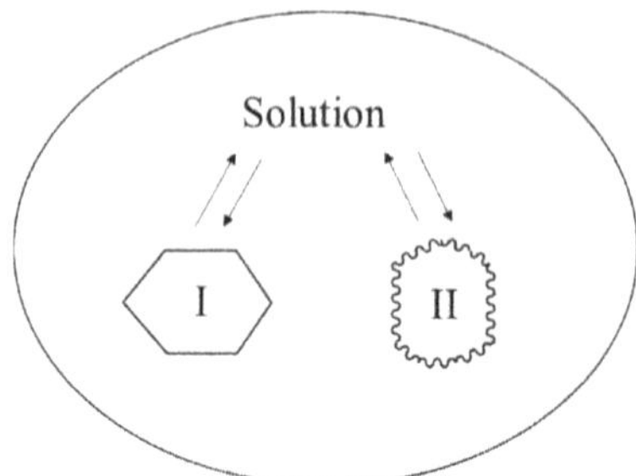

Figure 11. Thermodynamic principle of solution phase transition.

Thus, the lower the melting point of the crystalline polymorph has, generally, it shows the higher the solubility.

However, it cannot be judged so for the compounds that exhibit complex solid-solid phase transitions among the crystalline polymorphs. (cf. Figure 2.15 in the textbook.)

(iii)

G_I, G_{II} and G_S are set as the Gibbs energies of crystalline polymorph I, crystalline polymorph II and solution, respectively. From the explanation for (i) described above, when these two crystalline polymorphs have equal solubilities, it is the temperature of solution phase transition. Since both the crystal polymorphs I and II do not change their shape at this temperature, the crystalline polymorph I and the solution are in equilibrium, and the crystalline polymorph II and the solution are also in equilibrium. Therefore,

$$G_I = G_S = G_{II}$$

$$\therefore \Delta G = G_I - G_{II} = 0$$

Thus, at the point of intersection of the solubility curves (point of equal solubility), the difference in Gibbs energy between Polymorph I and Polymorph II is zero. Hence, it can be realized that a solid-to-solid phase transition occurs at this temperature.

Problem 9. Explain the composition phase diagrams Figs. 21 and 22 in the textbook.

Figures 12 and 13 drawn in below are the component phase diagrams of binary systems.

When we assume the molar fraction X_1, the melting enthalpy ΔH_{f1}, the freezing point T_1 (which can be used also as the melting point), and the gas constant R for Component 1 in the two-component system,

$$\ln X_1 = \frac{\Delta H_{f1}}{R}\left(\frac{1}{T_1} - \frac{1}{T}\right)$$

This formula holds also for Component 2,

$$\ln X_2 = \frac{\Delta H_{f2}}{R}\left(\frac{1}{T_2} - \frac{1}{T}\right)$$

$$\ln(1 - X_1) = \frac{\Delta H_{f2}}{R}\left(\frac{1}{T_2} - \frac{1}{T}\right) \quad (\because X_2 = 1 - X_1)$$

These two equations are called the **Le Chaterier-Schröder equation**, and the freezing point depression curve of the component phase diagram is obtained by taking X_1 on the horizontal axis and T on the vertical axis. The intersection point of the two curves is the eutectic point.

As can be seen from the **Le Chateriier-Schröder equation**, the shape of the curve is determined by the melting enthalpy and the melting point, and the number of the eutectic points are as many as the combination number between the freezing point depression curves in the right (Component 1) and left (Component 2) sides.

Figure 12 is a phase diagram of α and δ isomers of hexachlorocyclohexane. The δ isomer has two polymorphs with a different melting enthalpy and melting point, so that it gives two different freezing point depression curves. On the other hand, the α isomer shows no polymorphism, so there it gives only one freezing point depression curve. Therefore, since the number of eutectic points is the number of curves for δ isomer times the number of curves for the α isomer, *i.e.*, $2 \times 1 = 2$ eutectic points appear.

Figure 13 is a phase diagram of the γ and δ isomers of hexachlorocyclohexane. The γ and δ isomers have three and two crystalline polymorphs, respectively. Therefore, $3 \times 2 = 6$ eutectic points appear.

Problem 10. The following phase transition sequence was observed in a liquid crystalline compound. Above the arrows are the phase

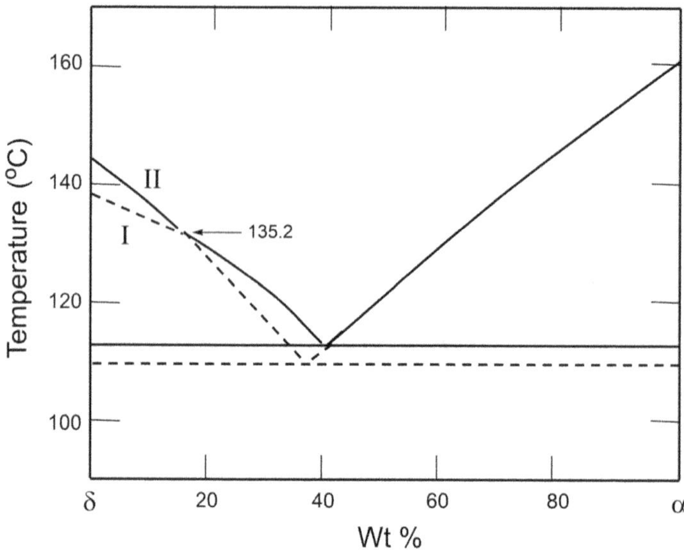

Figure 12. Composition diagram for the δ- and α-hexachlorocyclohexanes. The δ-isomer has two crystalline polymorphs.

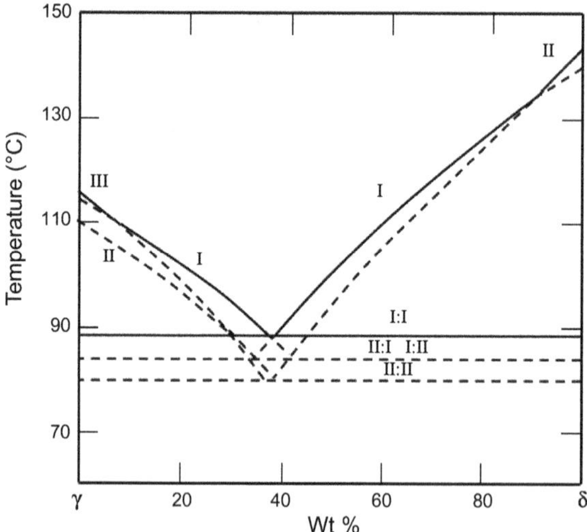

Figure 13. Composition diagram for the γ- and δ-hexachlorocyclohexanes. The γ- and δ-hexachlorocyclohexanes have three and two crystalline polymorphs, respectively.

transition temperatures and below the arrows are the phase transition enthalpy changes. K, M and L represent a crystalline phase, a liquid crystalline phase, and a liquid phase, respectively. Draw a schematic of the G–T diagram for this compound (cf. (ii-1-2)).

$$K_1 \xrightarrow[30\,\text{kJ/mol}]{30^\circ\text{C}} K_2 \xrightarrow[100\,\text{kJ/mol}]{50^\circ\text{C}} M_1 \xrightarrow[10\,\text{kJ/mol}]{150^\circ\text{C}} M_2 \xrightarrow[2\,\text{kJ/mol}]{180^\circ\text{C}} L$$

$$G = H - TS$$

Therefore,

$$\Delta G = \Delta H - T\Delta S$$

At the phase transition point T_t,

$$\Delta G = 0$$

Therefore,

$$0 = \Delta H_t - T_t \Delta S_t$$

$$\therefore \Delta S_t = \frac{\Delta H_t}{T_t}$$

From $\left(\frac{\partial G}{\partial T}\right)_P = -S$, the slope of each straight lines in the G–T diagram is $-S$, and the difference between the slopes of the two straight lines is ΔS.

Hereupon, if the entropy changes in $K_1 \to K_2$, $K_2 \to M_1$, $M_1 \to M_2$, and $M_2 \to L$ are set as ΔS_1, ΔS_2, ΔS_3 and ΔS_4, respectively,

$$\Delta S_1 = (30\,\text{kJ/mol}) \div (303\text{K}) = 99\,\text{J/mol}\cdot\text{K}$$

$$\Delta S_2 = (100\,\text{kJ/mol}) \div (323\text{K}) = 310\,\text{J/mol}\cdot\text{K}$$

$$\Delta S_3 = (10\,\text{kJ/mol}) \div (423\text{K}) = 24\,\text{J/mol}\cdot\text{K}$$

$$\Delta S_4 = (2\,\text{kJ/mol}) \div (453\text{K}) = 4.4\,\text{J/mol}\cdot\text{K}$$

Therefore, $\Delta S_2 > \Delta S_1 > \Delta S_3 > \Delta S_4$

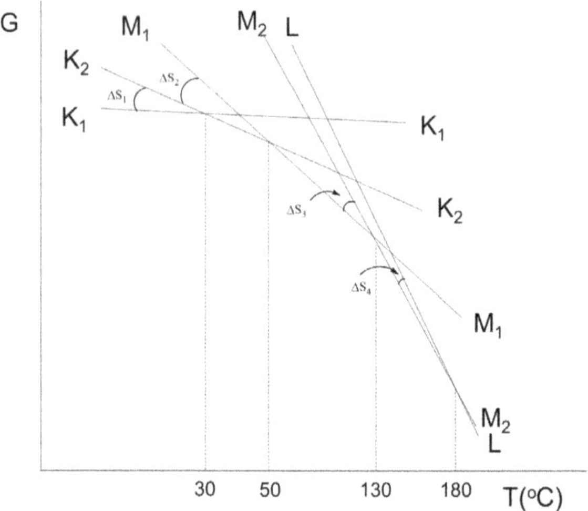

Figure 14. G–T diagram for the liquid crystalline material in Problem 10.

With this in mind, the G–T diagram of this liquid crystal material can be schematically drawn as shown in Figure 14.

Problem 11. The differential scanning calorimetry (DSC) was used to determine the phase transition enthalpy change and phase transition entropy of the liquid crystalline compound PAP as follows.

(DSC measurements)

Liquid crystalline compound used for the measurement: *p*-azoxydiphenetole, abbreviation: PAP, molecular weight: 286.23

Reference compound: phenanthrene, molecular weight: 178.23,

DSC measurement of 11.19 mg of the newly synthesized liquid crystalline compound PAP gave a thermogram (thermal analysis curve) as shown in B below.

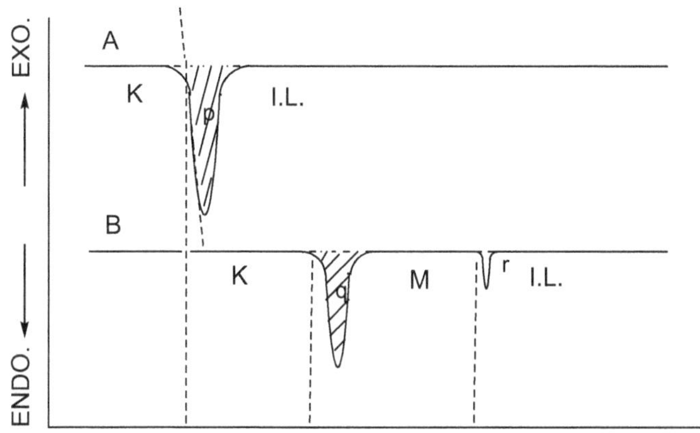

Figure 15. DSC thermograms.

The phase transition temperatures were determined by tangent intersection method and the following temperatures were obtained.

$$\text{K} \xrightarrow{137.8^\circ\text{C}} \text{M} \xrightarrow{167.8^\circ\text{C}} \text{I.L.}$$

Furthermore, since the transition enthalpy (ΔH) can also be determined from these thermograms, the following calculations were carried out.

(i) Calculation of phase transition enthalpy changes

Another DSC measurement was carried out using 18.08 mg of phenanthrene known the transition enthalpy change, under the same conditions as in the case of B. A thermogram could be obtained like A shown in the figure. The transition enthalpy changes at 100°C of phenanthrene is already known to be $\Delta \text{H} = 4.46\,\text{kcal/mol}$ from other methods.

In the present DSC method, the area of the hatched part p of the thermogram of A in the figure (and the weight of the paper cut off) is proportional to the enthalpy change ΔH. The weight of this paper was $\text{p} = 1987\,\text{mg}$. The weights of the paper cut from the thermogram of B in the figure were $\text{q} = 1125\,\text{mg}$ and $\text{r} = 86.99\,\text{mg}$, respectively.

By using the above data, determine the transition enthalpy (ΔHq, ΔHr) and transition entropy (ΔSq, ΔSr) of the liquid crystalline compound PAP.

(In the present modernized DSC measurements, the computer program enables us automatically calculate the phase transition enthalpies and entropies as if a black box. However, here, in order to learn the principle, it is calculated using the primitive method of the old days.)

(ii). Taking account of the magnitude of ΔS obtained in Problem 11-1, draw a schematic of the G–T diagram for this compound PAP (cf. Problem 10 and Text (ii-1-2)).

(i)

Since the molecular weight of phenanthrene is 178.23 g/mol, converting 18.08 mg to the amount of substance,

$$(18.08 \times 10^{-3} g) \div (178.23\, g/mol) = 1.014 \times 10^{-4}\, mol$$

Also, since the transition enthalpy per 1 mol of phenanthrene is 4.46 kcal/mol, the enthalpy measured by DSC this experiment is

$$\Delta H_p = 4.46 \frac{kcal}{mol} \times (1.014 \times 10^{-4}\, mol) = 0.452\, cal$$

This enthalpy of 0.452 *cal* is proportional to the mass p = 1987 mg of the paper cut from the thermogram.

The molecular weight of PAP is 286.23 g/mol, so 11.19 mg can be converted into the amount of substance

$$(11.19 \times 10^{-3} g) \div (286.23\, g/mol) = 3.909 \times 10^{-5}\, mol$$

The mass q = 1125 mg of paper cut from the thermogram at the K → M transition in PAP corresponds to the transition enthalpy, ΔH_q. Therefore.

$$0.452\, cal : 1987\,\text{mg} = \text{X}\, cal \;: 1125\,\text{mg}$$

$$X = \frac{0.425 \times 1125}{1987} = 0.256\, cal$$

This is the transition enthalpy at $3.909 \times 10^{-5}\, mol$, so the transition enthalpy per mol is

$$\Delta H_q = \left(\frac{0.256\, cal}{3.909 \times 10^{-5}\, mol} \right) = 6.55 \times 10^3 \frac{cal}{mol} = 6.55 \frac{kcal}{mol}$$

In the same way, the transition enthalpy ΔH_r of M $\rightarrow$ I.L. is determined.

Since the mass of the cut paper is r = 86.99 mg,

$$0.452\,\text{cal} : 1987\,\text{mg} = \text{Y}\,\text{cal} \ : 86.99\,\text{mg}$$

$$Y = \frac{0.425 \times 86.99}{1987} = 1.98\,cal$$

$$\Delta H_r = \left(\frac{1.98 \times 10^{-2}\,cal}{3.909 \times 10^{-5}\,mol}\right) = 0.507 \times 10^3 \frac{cal}{mol} = 0.507 \frac{kcal}{mol}$$

Entropy is $\Delta S_t = \frac{\Delta H_t}{T_t}$ as obtained in Problem 10, so

$$\Delta S_q = \frac{\Delta H_q}{T_q} = \frac{6.55 \times 10^3 \frac{cal}{mol}}{410\,K} = 16.0 \frac{cal}{K \cdot mol}$$

$$\Delta S_r = \frac{\Delta H_r}{T_r} = \frac{507 \frac{cal}{mol}}{440.8\,K} = 1.15 \frac{cal}{K \cdot mol}$$

In summary, the above are

$$\Delta H_q = 6.55 \frac{kcal}{mol} \quad \Delta S_q = 16.0 \frac{cal}{K \cdot mol}$$

$$\Delta H_r = 0.507 \frac{kcal}{mol} \quad \Delta S_r = 1.15 \frac{cal}{K \cdot mol}$$

(ii)

Considering the same as in Problem 10, the G–T diagram for this compound can be drawn schematically as shown in Figure 16

Problem 12. The liquid crystalline phase M of PAP in Problem 11 is wanted to identify by miscibility test using polarizing microscopic observation. First, the microscopic observation revealed that this PAP exhibited a Schlieren texture, which includes a nematic phase and smectic C, F and I phases. Since there are two and four brushes in the Schlieren texture of this PAP, the possibility of the nematic phase is high. Therefore, in order to identify the identification more surely, a miscibility test was carried out using DSC between this

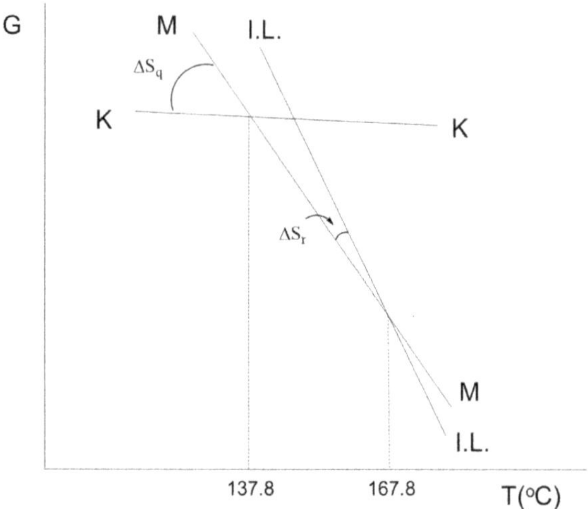

Figure 16. The G–T diagram for *p*-azoxydiphenetole (PAP).

unknown liquid crystalline phase-showing PAP and the known standard nematic-phase-showing compound PAA shown below.

CH_3O—C_6H_4—N=N(→O)—C_6H_4—OCH_3

p-Azoxydianisole: abbreviation PAA; molecular weight = 258.28

Here, if the M phase of PAP is immiscible with the standard N phase of PAA, the binary phase diagrams may become like as Figure 17[A] shown below, and we can conclude that it is a liquid crystal phase different from the N phase. If it mixes completely, the binary phase diagram may become like as Figure 17[B], and we can conclude that it is the same nematic phase as the N phase of PAA.

The unknown liquid crystalline phase-showing PAP was well mixed with this standard nematic-phase-showing PAA to prepare the plural DSC samples with changing the mixture ratio. The contents of PAA were listed in mol% in Table 2. The DSC of each sample was measured to obtain a thermogram. The representative DSC thermograms are shown in Figure 18. These transition temperatures were determined by tangent intersection method and are summarized also in Table 2.

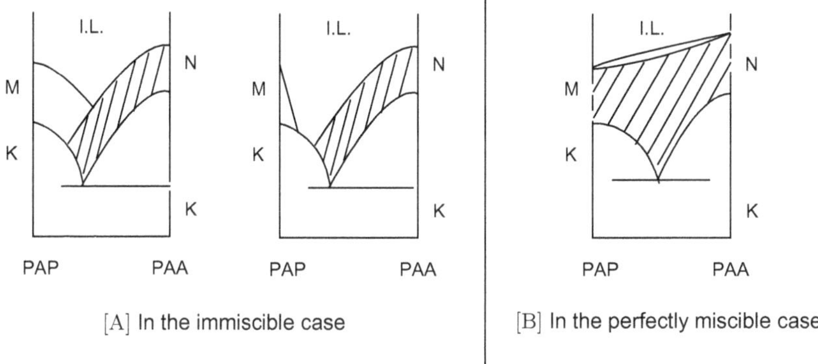

Figure 17. Miscibility test.

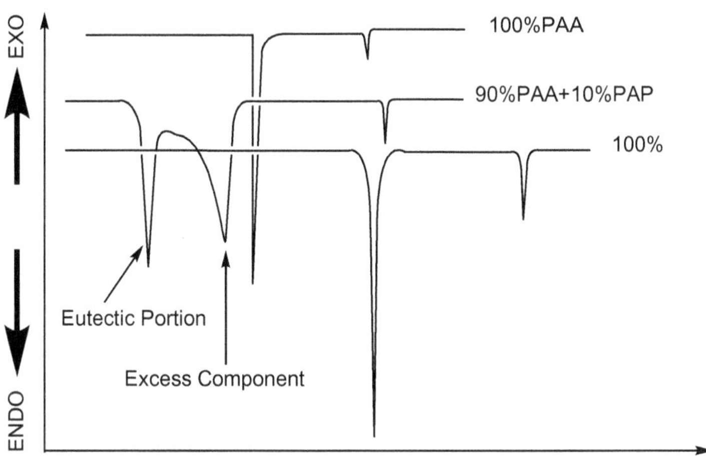

Figure 18. The representative DSC thermograms of the single component and the binary mixture.

Using these data, draw a two-component phase diagram and show that the PAP M phase and the PAA N phase are completely mixed. Also, find the experimental eutectic point from this phase diagram.

When we draw a binary phase diagram using the values in Table 2, Figure 19 can be obtained.

Since this figure looks just like Figure 17[B], it can be seen that the unknown liquid crystal phase M of PAP is completely mixed with the nematic phase N of PAA, the standard material. Therefore, it can

Table 2. Results of miscibility test between PAA and PAP.

Mol% of PAA	Transition point (°C)	Transition point (°C)	Transition point (°C)
0	—	137.8	167.8
4.5	97.0	134.7	166.6
7.3	97.0	133.3	165.5
14.7	97.3	129.0	163.1
24.6	97.3	123.0	159.7
34.9	97.4	116.2	156.3
42.2	97.4	111.1	153.3
51.3	97.4	104.6	151.1
57.5	97.4	100.3	148.8
60.0	97.4	—	147.5
65.6	97.4	100.2	145.5
73.3	97.4	104.6	143.3
80.2	97.4	108.4	141.2
85.4	97.4	111.3	139.8
90.2	97.4	113.4	138.7
93.3	97.4	115.1	137.7
100.0	—	119.5	136.5

be concluded from this miscibility test that the liquid crystal phase M of PAP is the nematic phase N. This is one of the identification methods of the liquid crystal phase.

It can be also seen from Figure 19 that the eutectic point exists at 97.4°C for 60.0% PAA.

Problem 13. (i) When we assume that a saturated solution contains Solute 1 at a molar fraction x_1. In the saturated solution, the dissolved and insoluble solutes are in equilibrium. Therefore, the chemical potential $\mu_1^*(s)$ of Solute 1 in pure solid and the chemical potential μ_1 of Solute 1 in the solution are equal.

$$\therefore \mu_1^*(s) = \mu_1$$

The chemical potential of Component 1 in a solution is given by using the pure liquid chemical potential $\mu_1^*(\ell)$ of as

$$\mu_1 = \mu_1^*(l) + RT \ln x_1$$

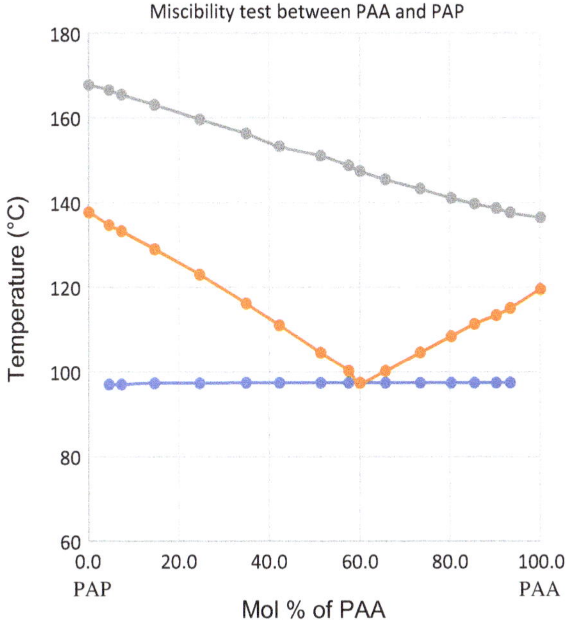

Figure 19. Binary phase diagram of PAP and PAA.

can be written. From $\mu_1^*(s) = \mu_1$,

$$\mu_1^*(s) = \mu_1^*(\ell) + RT \ln x_1$$

$$\therefore \ln x_1 = \frac{\mu_1^*(s) - \mu_1^*(\ell)}{RT} = -\frac{\Delta G_{f1}}{RT}$$

can be derived, where ΔG_{f1} is the Gibbs energy of solidification (freezing) of the pure solvent (Component 1). To find the relationship between the change in composition and the change of freezing point (= melting point), we differentiate both sides with respect to temperature,

$$\frac{d(\ln x_1)}{dT} = -\frac{1}{R}\frac{d(\Delta G_{f1}/T)}{dT}$$

Here, from the Gibbs-Helmholtz equation $\left\{\frac{\partial(G/T)}{\partial T}\right\}_P = -\frac{H}{T^2}$,

$$\therefore \frac{d(\ln x_1)}{dT} = \frac{\Delta H_{f1}}{RT^2}$$

When this is changed into a separated variable form,

$$d(\ln x_1) = \frac{\Delta H_{f1}}{RT^2} dT$$

can be derived. When we consider to integrate both the sides from $\ln x_1 = 0$ to $\ln x_1 = \ln x_1$.

The $\ln x_1 = 0$ corresponds when $x_1 = 1$, and when $x_1 = 1$, it corresponds to the melting point of pure Component 1: $T = T_1$. On the other hand, when $\ln x_1 = \ln x_1$, it corresponds to an arbitrary melting point of Component 1: $T = T$.

Thus, the integration range is $T = T_1 \sim T$, when $\ln x_1 = 0 \sim \ln x_1$. Therefore,

$$\int_0^{\ln x_1} d(\ln x_1) = \frac{1}{R} \int_{T_1}^{T} \frac{\Delta H_{f1}}{T^2} dT$$

Hereupon, if the melting enthalpy of Component 1 is constant over the temperature range of interest,

$$\int_0^{\ln x_1} d(\ln x_1) = \frac{\Delta H_{f1}}{R} \int_{T_1}^{T} \frac{1}{T^2} dT$$

$$\therefore \ln x_1 = \frac{\Delta H_{f1}}{R} \left(\frac{1}{T_1} - \frac{1}{T} \right) \qquad \text{(Equation 1)}$$

Similarly for Component 2,

$$\ln x_2 = \frac{\Delta H_{f2}}{R} \left(\frac{1}{T_2} - \frac{1}{T} \right) \qquad \text{(Equation 2)}$$

From $x_1 + x_2 = 1$,

$$\therefore \ln (1 - x_1) = \frac{\Delta H_{f2}}{R} \left(\frac{1}{T_2} - \frac{1}{T} \right) \qquad \text{(Equation 2')}$$

As described above, Le Chaterier-Schröder's Equations (1) and (2) can be theoretically derived.

Table 3. Calculation results by Equations (13-1) and (13-2).

x_1	T(1)K	T(2)K	T(1)°C	T(2)°C
0.0	—	411.0	—	137.9
0.1	314.5	405.7	41.4	132.5
0.2	334.6	399.9	61.4	126.7
0.3	347.5	393.5	74.4	120.3
0.4	357.4	386.4	84.2	113.2
0.5	365.4	378.3	92.2	105.2
0.6	372.2	368.9	99.0	95.7
0.7	378.1	357.4	105.0	84.2
0.8	383.5	342.3	110.3	69.2
0.9	388.3	319.3	115.1	46.2
1.0	392.7	—	119.6	—

T(1) = $1/(1/392.7 - \ln(x_1)/3638)$
T(2) = $1/(1/411 - \ln(1 - x_1)/3296)$

(ii) Into Equation 1 derived above, we substitute the values, $\Delta H_{f1} = 7.23\, kcal/mol$ and $T_1 = 119.5°\text{C} + 273.15 = 392.7\text{K}$, to obtain

$$\ln x_1 = \frac{7.23 \times 10^3\, cal/mol \times 4.184\, J/cal}{8.314\, J/K \cdot mol}\left(\frac{1}{392.7K} - \frac{1}{T}\right)$$

When we solve this equation for T,

$$T = \left(\frac{1}{392.7K} - \frac{\ln x_1}{3638K}\right)^{-1} \cdots\cdots (13\text{-}1)$$

On the other hand, into Equation 2′, we substitute the values, $\Delta H_{f2} = 6.55\, kcal/mol$, and $T_2 = 137.8°\text{C} + 273.15 = 411.0\text{K}$, to obtain

$$\ln(1 - x_1) = \frac{6.55 \times 10^3\, cal/mol \times 4.184\, J/cal}{8.314\, J/K \cdot mol}\left(\frac{1}{411.0\, K} - \frac{1}{T}\right)$$

$$T = \left(\frac{1}{411.0K} - \frac{\ln(1 - x_1)}{3296K}\right)^{-1} \cdots\cdots (13\text{-}2)$$

When we calculate these formulas (13-1) and (13-2) by using the spreadsheet software Excel, the following Table 3 and Figure 20 can be obtained.

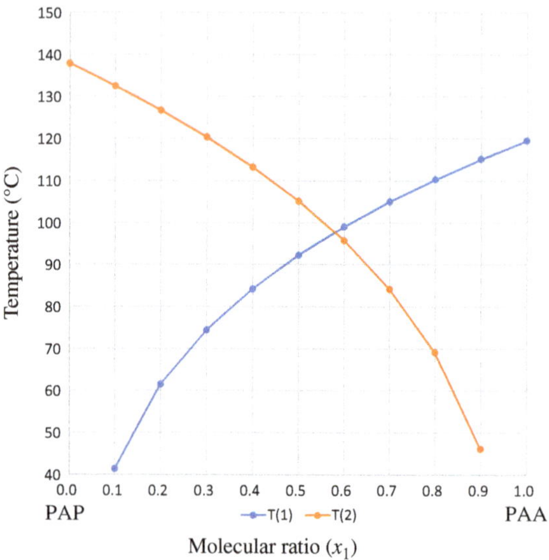

Figure 20. Freezing point depression curves of PAP and PAA calculated from the Le Chaterier-Schröder equations.

From the intersection in this figure, it can be seen that the theoretical eutectic point is at PAA $= 58.0\,\text{mol\%}$ and 98.0°C. On the other hand, the measured value of the eutectic point was at PAA $= 60.0\,\text{mol\%}$ and 97.4°C, as shown in Figure 19 of Problem 12, so it can be seen that this theoretical value is almost the same as the measured value.

Problem 14. Explain how to prepare metastable crystal polymorphs, taking HMX(I) to HMX(IV) as an example.

(1) Place the desired polymorph in the most stable temperature range. When the crystals are wetted with a solvent, the transformation can quickly proceed by a solution phase transition. It is important to know the phase transition temperature among the polymorphs in detail in order to perform the solution phase transition in the stable range of the desired polymorph.

(Example)
Looking at the G–T diagram for the HMX polymorph shown in Problem 7, the most stable temperature range for the HMX(IV) polymorph is between 165.5°C $< T <$ T_{m4}. Therefore, when the HMX(I) polymorph is placed and maintained in this temperature range with a small amount of poor solvent, it undergoes a solution phase transition to Polymorph IV. It is quenched (rapidly cooled) to room temperature and the solvent is removed quickly enough to prevent relaxation to other polymorphs to give pure Polymorph IV.

(2) When the melt (I.L.) is supercooled by quenching, a metastable polymorph is resulted before the appearance of the stable form.

(cf. Already noted in Problem 7.)

(3) When we recrystallize the solutions with various concentrations and cooling rates, a variety of different polymorphs can be obtained, depending on the concentration and the cooling rate. It is necessary to obtain less stable polymorphs that we use more concentrated hot solutions and carry out faster quenching.

Table 4. Preparation methods of various polymorphs of HMX from a saturated acetone solution of HMX(I) (free of solids).

Target polymorph	Solution	Cooling method
HMX(I)	200 mℓ of hot saturated acetone solution	Spontaneously cooled to room temperature
HMX(II)	50 mℓ of hot saturated acetone solution	Cooled to room temperature with gentle stirring
HMX(III)	30 mℓ of hot saturated acetone solution	Cooled in a cold water bath with stirring
HMX(IV)	5 mℓ of acetone solution	Poured directly onto crushed ice
HMX(IV)	5 mℓ of acetone solution in a test tube	Placed in a dry ice-ethanol bath with shaking

(Example)
However, since the crystals may transform by the solution phase transition to the stable HMX(I) polymorph at the temperature higher than 115.5°C, it is necessary to quickly filter off the crystals to remove the remained solution in all the cases.

(4) When the crystals are heated to sublime and deposited on a cold cover glass, metastable crystals are obtained. The lower the temperature of the coverslip becomes, the more unstable polymorphs can be obtained.

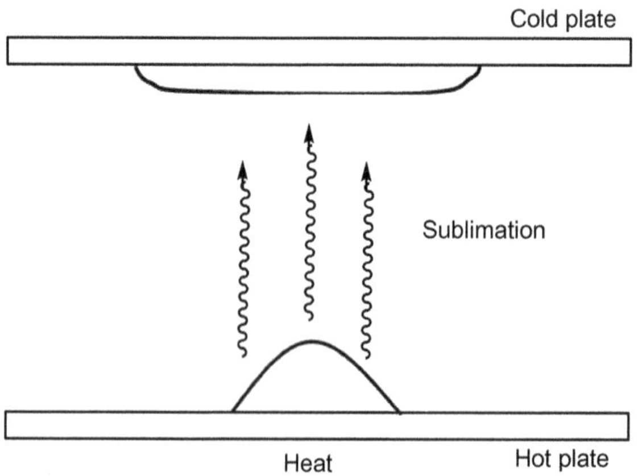

Figure 21. Formation of a crystalline polymorph from the vapor by sublimation. The lower the temperature of the coverslip becomes, the more unstable polymorphs can be obtained.

Problem 15. K (crystal) → K (crystal) phase transition is easily superheated, whereas K (crystal) → M (liquid crystal) phase transition not superheated. Why?

In the crystalline phase, the molecules are tightly aggregated, and the rearrangement of molecules is difficult to occur. A large activation energy mountain (ΔE_a) must be climbed over to cause a crystal–crystal phase transition. Therefore, a large amount of energy must be applied from the outside to cross this mountain and change the arrangement of the molecules. For this reason, the

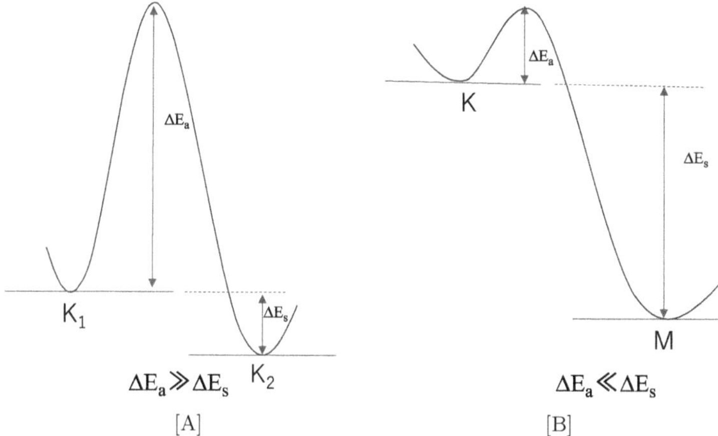

Figure 22. Activation energy (ΔE_a) and stabilization energy (ΔE_s) in [A] crystal phase (K_1) — crystal phase (K_2) transition and [B] crystal phase (K) — liquid crystal (M) phase transition.

crystal–crystal phase transition is prone to superheating. However, the stabilization energy (ΔE_s) at the crystal–crystal phase transition is generally much smaller than this activation energy ($\Delta E_a \gg \Delta E_s$: see Fig. 18[A]).

On the other hand, in the liquid phase and the liquid crystal phase, the molecules are relatively loosely aggregated and are in a state where the rearrangement of molecules easily occurs. A relatively small activation energy peak (ΔE_a) only have to be climbed over to cause the crystal-to-liquid crystal phase transition, and the rearrangement of molecules will more easily occur. Therefore, only by applying a small amount of energy from the outside, this mountain can be easily crossed to rearrange the molecules. Accordingly, the crystal-liquid crystal phase transition is less likely to be superheated. However, the stabilization energy (ΔE_s) in the crystal-liquid crystal phase transition is generally much larger than this activation energy ($\Delta E_a \ll \Delta E_s$: see Fig. 18[B]).

Problem 16. Describe the verification methods, at least two, whether polymorphs exist.

We can verify whether polymorphism exists or not, by using a polarizing microscope equipped with a hot stage.

(1) Once the given compound is completely melted and then turned back to cool. If the melt spontaneously recrystallizes and then a solid–solid phase transition occurs on further cooling, the compound has polymorphs.

$$\text{I.L.} \xrightarrow{\text{cooling}} K_1 \xrightarrow{\text{cooling}} K_2$$

(2) If a solid–solid transition occurs on a heating stage, the compound has polymorphs. The larger the sample becomes, the higher the possibility of transformation becomes.

$$K_1 \xrightarrow{\text{heating}} K_2$$

(3) The compound is sublimated and deposited on a glass plate at a low temperature in order to obtain a sublimate (cf. Figure 21 in Chapter 2 in this book). The sublimate is mixed with the original sample and then the mixture is put into a saturated solution of both the sublimate and the original sample. (i) If the two are polymorphs, we can observe that the stable polymorph is less soluble and the metastable polymorph is more soluble, and that a solution phase transition occurs in which the metastable polymorph dissolves while the stable polymorph grows (cf. Figure 9 in Chapter 2 in this book). This continues until the metastable polymorph completely transforms to the stable polymorph. (ii) If the two are different compounds, one dissolves but another one does not grow. (iii) If the two are the same polymorph of the same compound, nothing changes.

(4) An excess of crystalline compound is placed in a small volume of solvent, and the temperature is kept just below the melting point of the compound and is held for several hours. Then the crystals are quickly isolated. The product has likely transformed to a higher temperature polymorph. Hereupon, it is examined whether a solution phase transition occurs between this product

and the original compound in the same manner as in the above (3).

(5) When a small amount of solution is rapidly cooled (quenched) to recrystallize the compound, it is likely to give a metastable polymorph. Hereupon, it is examined whether the solution phase transition occurs between this product and the original compound in the same manner as in the above (3).

Problem 17. Explain more than three methods to prove whether two samples are crystal polymorphs of the same compound. Be sure to mention double melting behavior.

(a) If two crystalline samples have different crystal forms, *i.e.*, axial ratios, refractive indices, densities, X-ray powder patterns, and if they undergo a solid-solid phase transition or a solution phase transition with each other, they are the polymorphs of the same compound.

(b) We heat a mixture of two kinds of crystals (A + B) to the first lower melting point to melt, and then keep the resulting liquid at this temperature. If the liquid completely recrystallizes by using the remained crystals (B) as the seeds and they melt completely at the melting point of B on further heating, the two crystalline forms are polymorphs of the same compound.

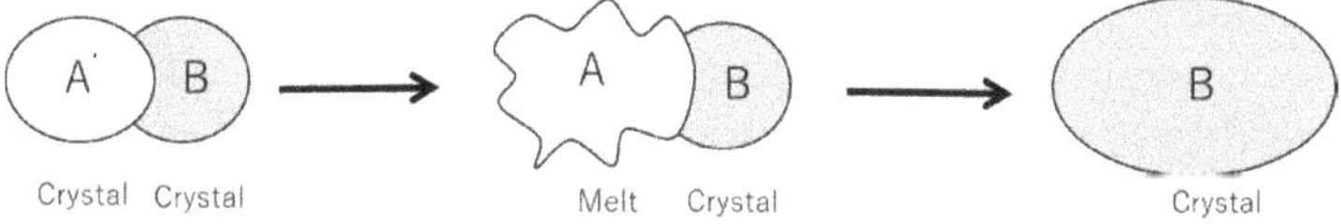

(c) If the melts of the two samples show exactly the same properties (refractive index, temperature dependence of refractive index), and the crystallinity of the two is different like as the above (a), they are the polymorphs of the same compound.

(d) We heat one of the two crystalline samples A and B to melt; the melt is supercooled slightly below its melting point; onto the supercooled liquid, two kinds of crystalline samples, A and B, are inoculated at two different locations; each crystal grows and collides with each other to form boundaries; we observe this boundary with a polarizing microscope. If one crystal B continues to grow through another crystal A, like as swallowing, these two crystalline samples A and B are polymorphs of the same compound.

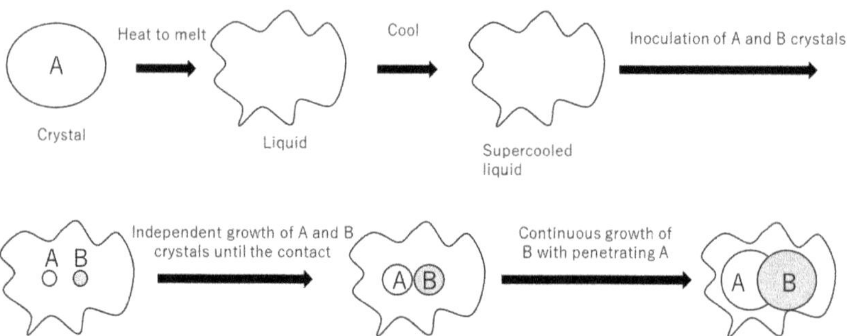

(e) If a solution phase transition occurs between two samples, they are polymorphs of the same compound (cf. Figure 11 of Chapter 2 in this book).

(f) (Case I) We bring two crystalline samples, A and B, close to each other and heat them. If one of them, A, undergoes a solid-solid phase transition and then it melts at the same temperature as B, it is highly likely that the two samples are polymorphs of the same compound. However, it's not proof. Because they may happen to have the same melting point, although they are different compounds.

(Case II) We bring two crystalline samples, A and B, close to each other and heat them. If one of them, A, melts, and this melt is completely recrystallized by using another crystal B as the seeds, and it melts completely at one temperature on further heating, these two samples are polymorphs of the same compound; *i.e.*, if two samples exhibit **double melting behavior**, they are polymorphs of the same compound.

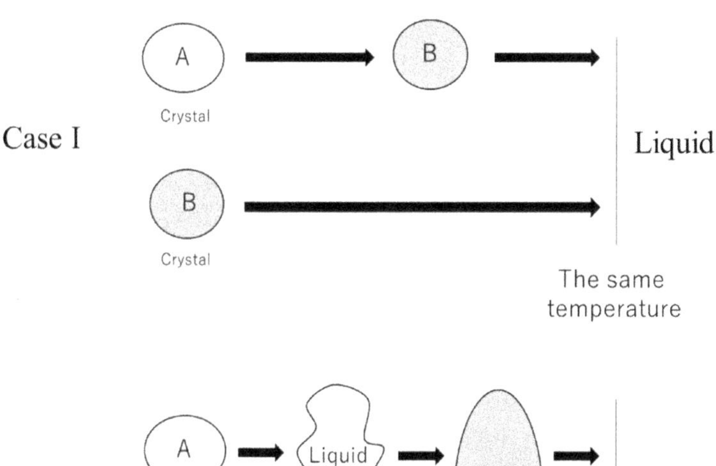

(g) If two well-mixed samples (A+B) are sublimated to give one sublimate C, the two samples, A and B, are polymorphs of the same compound.

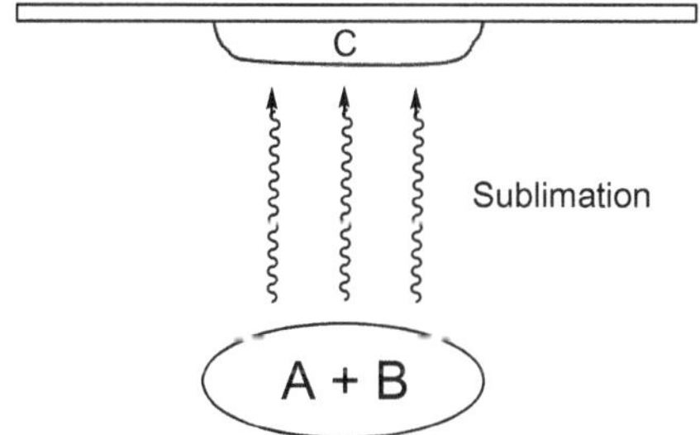

Problem 18. Discuss the application of polymorphism. Be sure to mention how to prepare suppositories and quick efficacy insulin.

(1) A suppository is a solid product made by mixing medicine with theobroma oil and molding it into a rocket shape. If the patient is a

baby or infant who cannot take the drug orally, the suppository is inserted through the anus so that the patient can take the drug.

The base of the suppository is theobroma oil which is extracted from cacao and the raw material of chocolate. Theobroma oil has six crystalline polymorphs, each with a different melting point. The lowest melting point is Polymorph I at 17.3°C and the highest melting point is Polymorph VI at 36.3°C. The eatable chocolates and medical suppositories are used Polymorph V with a melting point of 33.8°C. The inside of the mouth and anus is just 33–34°C, so that chocolates and suppositories melt well in the body.

If we fail to recrystallize and solidify it at 27°C, it will melt just by touching it with a finger. In this case, we solidify it again at 33°C to form Polymorph V: in order to make suppositories, we should melt theobroma oil at 60–70°C, and then mix medicine such as antipyretic; this melt is poured this into a rocket-shaped mold kept at 33°C to solidify it. On the other hand, if such suppositories are inadvertently left in a warm room for a long time, it transforms into the most stable Polymorph VI that will not melt in the mouth or anus. That is why, when you get a suppository from a pharmacy, you are instructed to put it in the refrigerator immediately. If the suppository becomes the most stable Polymorph VI, it will not melt in the body; it will be a disaster with a continuously crying baby. Parents with a baby or young infant should be very careful when storing suppositories.

In this way, suppositories take advantage of solid polymorphism.

(2) How to make fast-acting insulin and slow-acting insulin

The duration of insulin action is controlled by crystallinity. Insulin is used as a zinc complex by reacting with zinc chloride.

- Fast-acting insulin: made of amorphous material and provided as the fine ground powder.
- Slow-acting insulin: 70% is crystalline and 30% is amorphous.

Amorphous state refers to the glassy state or supercooled liquid, which is more unstable than the most unstable metastable crystalline polymorph and exhibits a solubility higher than the most unstable metastable crystalline polymorph.

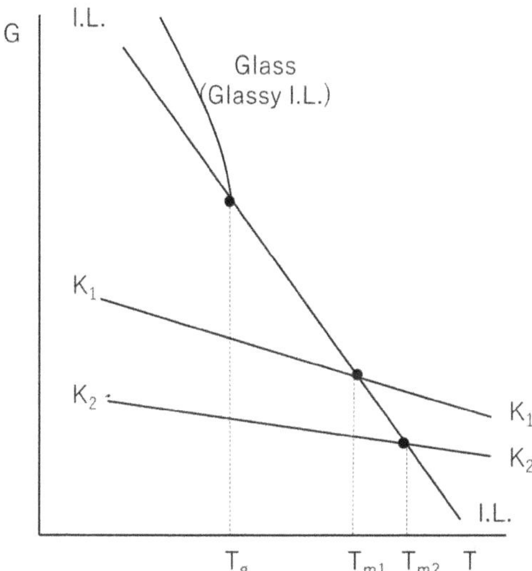

Figure 23. A schematic G–T diagram of a glassy liquid (Glassy IL) originated from an isotropic liquid (IL).

Figure 23 shows a schematic G–T diagram of a glassy isotropic liquid (Glassy I.L.) generated from an isotropic liquid (I.L.). When the I.L. is rapidly cooled (=quenched), the supercooled isotropic liquid solidifies below the glass transition temperature (T_g) without the crystallization. As can be seen from this G–T diagram, Glassy I.L is much higher than I.L., so that it is much more unstable than I.L. and much more soluble than the polymorphs K_1 and K_2. Therefore, the amorphous insulin has an extremely high solubility at room temperature, and its concentration in blood rises instantaneously, to show the fast-acting. The crystalline polymorphs K_1 and K_2 are much more stable than Glassy I.L., so that both crystalline polymorphs are much less soluble at room temperature and dissolve slowly in small increments to show the slow-acting.

Severely diabetic patients must inject insulin several times a day. If the insulin is not fast-acting but slow acting, the patient is at risk of life. Therefore, if the fast-acting insulin is so poorly stored that it transforms into the slow-acting insulin, it is a big problem of life and death. As can be seen from the discussion of Figure 23 mentioned above, the fast-acting insulin must be stored as a fine ground powder.

If it is stored in the large size without fine powdering, they will rapidly transform into the stable crystalline polymorphs.

However, long ago, an American pharmaceutical company produced a large amount of the fast-acting insulin at once and stored it in a large size for a long time, as they are; just before shipment they ground it to the fine powder, and shipped it. The insulin caused dozens of deaths across the United States. The pharmaceutical company had to pay a large amount of compensation to the bereaved families. Therefore, the fast-acting insulin must be stored as a very fine powder just after producing. If stored in a large size, it rapidly transforms into the stable crystalline polymorphs with low efficacy and slow-acting. Pharmaceutically, fine powdering is a very important storage technology.

From the example of insulin described above, it can be seen that pharmaceutical companies must have a good knowledge of not only single-molecule synthesis technology but also molecular assembly control technology.

Problem 19. Read the following newspaper article and comment on this lawsuit from the viewpoint of polymorphism and drug efficacy.

December 26, 2013 Nikkei newspaper morning issue
Nissan Chemical sued for the patent infringement to her hyperlipidemia drug

On the 25th, Nissan Chemical Industries, Ltd., which manufactures the drug for hyperlipidemia treatment "Livalo," sued Tokyo District Court for the manufacturing and sales of seven generic medicines such as Towa Pharmaceutical and Daito, for infringing the Livalo patent. They are seeking a suspension of production and sales.

According to the lawsuit, these seven companies have infringed a patent on the crystalline form of the active ingredient of Livalo owned by Nissan Chemical. These seven companies began selling the corresponding generic medicines after receiving a drug price listing on March 13 because the patent for the compound was expired in August of this year. However, Nissan Chemical claims that the patent for the crystal form of this compound is still valid until 2024.

Medicinal efficacy is, of course, related to its specific molecular structure, but is also highly dependent on its crystalline polymorph. In general, the medical efficacy of a compound becomes higher for the crystalline polymorph with higher solubility at room temperature; it becomes lower for the crystalline polymorph with lower solubility at room temperature. This is because when we take a polymorph that causes the concentration in blood the highest, it gives the highest medical efficacy.

With this in mind, we will read this newspaper article.

For the hyperlipidemia drug "Rivalo," Nissan Chemical Industries has two patents with different expiration periods: a patent concerning the specific molecular structure until 2013, and another patent concerning the preparation of the highest soluble crystalline polymorph until 2024.

According to the present newspaper article, the patent on the molecular structure of "Livalo" has expired, but the patent on the manufacturing technology of the crystal polymorph showing the highest medical efficacy has not yet expired. Despite this, seven generic pharmaceutical manufacturers including Towa Pharmaceutical, Daito, and so on, manufactured and marketed this drug with the same molecular structure and the same crystalline polymorph, so that Nissan Chemical brought into court for the patent infringement.

Chapter 3

Solutions to End-of-Chapter Problems

Problem 1. Describe the characteristics of liquid crystal and plastic crystal, respectively, and explain the difference.

Conventionally, it is generally accepted that the state of matter can be divided into three states, but this include neither liquid crystals nor plastic crystals, which is an error in modern science. The conception of the new state of matter is shown in Figure 1. As can be seen from this figure, both liquid crystals and plastic crystals are intermediate states between solid and liquid, and each has different characteristics as described in the followings.

(Characteristics of liquid crystals)

When a certain kind of compound is heated, the crystal first melts at T_1(K) to become a cloudy viscous liquid, and on further heating it becomes a clear liquid at $T_2(>T_1)$. The liquid generated between T_1 and T_2 exhibits birefringence (optical anisotropy) under a polarizing microscope, like as crystals. Accordingly, this state was named liquid crystal, meaning a crystal with fluidity.

In liquid crystals, there is no three-dimensional space lattice like in crystals. Major characteristics of liquid crystals are fluidity or plasticity originated from disappearance of the long-range order of molecular gravity centers, and optical anisotropy due to the partially remaining order of molecular orientation. (Figure 2)

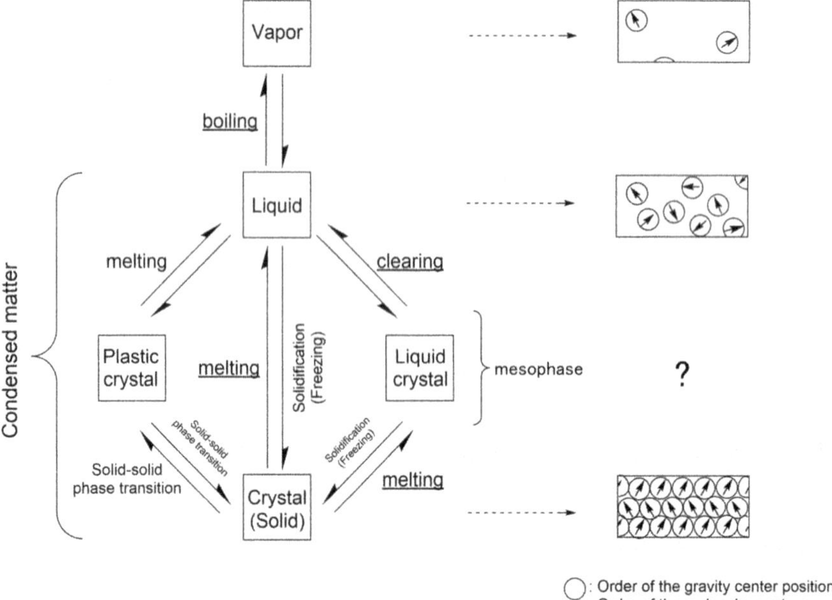

Figure 1. Novel recognition of matter states.

Note 1: However, some liquid crystals do not self-flow. They are soft and highly plastic when pressed.

Note 2: The "hard" liquid crystalline phases like as $S_G \sim S_L$, which have a very short-range three-dimensional space lattice (three-dimensionality).

(Conditions to judge liquid crystals)

(1) Most of the molecules are rod-like or disk-like with long-chain alkyl groups in the periphery.
(2) Both birefringence and fluidity (or plasticity) exhibit at the same time.
(3) The X-ray diffraction patterns do not show a long-range three-dimensional space lattice.

(Characteristics of plastic crystals)

As can be seen in the crystalline phase just below the melting point of carbon tetrachloride, cyclohexanol and so on, the phase gives

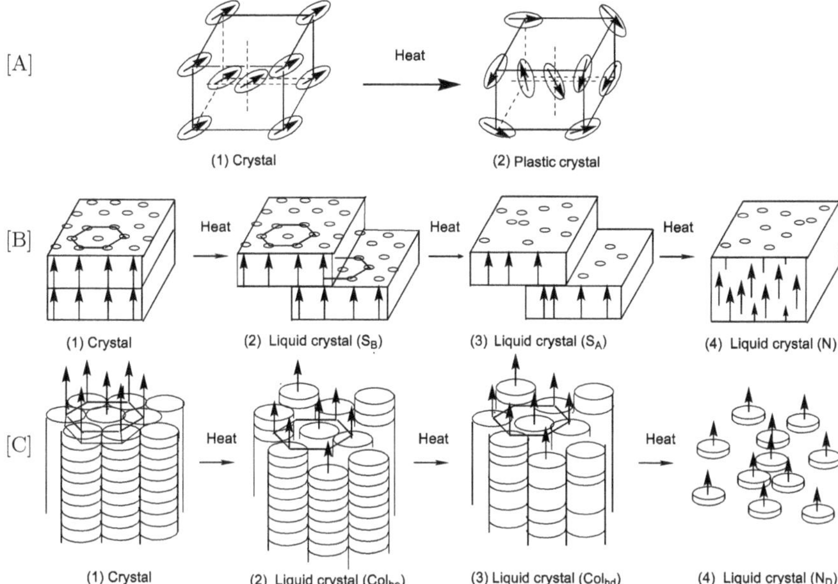

Figure 2. Stepwise disintegration of a three-dimensional lattice from the viewpoint of the long-range three-dimensional order of gravity center position and the long-range three-dimensional order of molecular posture: [A] in the case of spherical molecules, [B] in the case of rod-like molecules, and [C] in the case of disk-like molecules.

X-ray Laue spots and satisfies the image of a crystal in the sense that it has a clear melting point. Nevertheless, it is so far from the image of crystals, from a mechanical point of view, that it shows very high plasticity; in the extreme cases, it flows under its own weight. (Figure 2)

(**Conditions to judge plastic crystals**)

(1) The molecular structure should be nearly spherical.
(2) The melting entropy change ΔS_m must be small; the upper limit is approximately 21 J/K·mol as an upper limit: Timmermans' empirical rule.

Figure 1 schematically illustrates the positions of plastic crystal and liquid crystal among the three states of matter.

Table 1. Characteristics of the intermediate states: the difference between plastic crystal and liquid crystal.

	State \ Order	Long-distance 3D order of gravity center position	Long-distance 3D order of molecular posture
	Liquid	×	×
Intermediate states (mesophases)	Plastic crystal	○	×
Intermediate states (mesophases)	Liquid crystal	× ← Stepwise disintegration △	○
	Crystal	○	○

X: this mark means that the order does NOT exist. O: this mark means that the order exists.

Figure 2 illustrates stepwise disintegration of a three-dimensional lattice from the viewpoint of the long-range three-dimensional order of gravity center position and the long-range three-dimensional order of molecular posture: [A] in the case of spherical molecules, the order of order of molecular posture is first collapsed while maintaining the three-dimensional crystal lattice. [B] and [C] in the case of rod-like and disk-like molecules, the long-range order of molecular center of gravity is first stepwise collapsed while maintaining the order of molecular posture.

As can be seen from Figures 2[B] and 2[C], it can be seen that there is not only one type of liquid crystal phase but multiple types.

Hereupon, the characteristics of crystals, plastic crystals, and liquid crystals are summarized in Table 1, based on the long-range three-dimensional order of gravity center position and the long-range three-dimensional order of molecular posture.

As can be seen from this table, the melting of an (ordered) crystal into an (isotropic) liquid has two aspects: the stepwise disintegration of the three-dimensional crystal lattice and the collapse of the molecular posture (orientation).

When the molecule is nearly spherical, as shown in Figure 2[A], the barrier against molecular rotation within the crystal is small; the molecular orientation is disrupted at first while the three-dimensional crystal lattice is maintained. In this case a plastic crystal is realized.

Table 2. Types of liquid crystals.

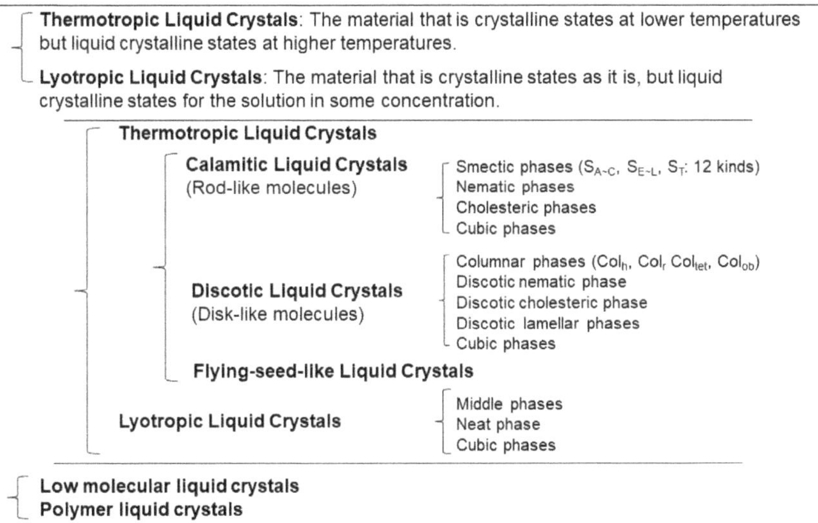

- **Thermotropic Liquid Crystals**: The material that is crystalline states at lower temperatures but liquid crystalline states at higher temperatures.
- **Lyotropic Liquid Crystals**: The material that is crystalline states as it is, but liquid crystalline states for the solution in some concentration.

- **Thermotropic Liquid Crystals**
  - **Calamitic Liquid Crystals** (Rod-like molecules)
    - Smectic phases ($S_{A\sim C}$, $S_{E\sim L}$, S_T: 12 kinds)
    - Nematic phases
    - Cholesteric phases
    - Cubic phases
  - **Discotic Liquid Crystals** (Disk-like molecules)
    - Columnar phases (Col_h, Col_r, Col_{tet}, Col_{ob})
    - Discotic nematic phase
    - Discotic cholesteric phase
    - Discotic lamellar phases
    - Cubic phases
  - **Flying-seed-like Liquid Crystals**
- **Lyotropic Liquid Crystals**
  - Middle phases
  - Neat phase
  - Cubic phases

- **Low molecular liquid crystals**
- **Polymer liquid crystals**

In contrast, when the molecules are rod-like or disk-like, as shown in Figures 2[B] and 2[C], they have strong intermolecular forces aligning them parallel or face-to face to each other; the molecular orientation is maintained, while the stepwise disintegration of the three-dimensional crystal lattice occurs to give multiple liquid crystalline phases.

Therefore, the two intermediate states of plastic crystal and liquid crystal are originated due to the separate disintegration of the two orders, *i.e.*, the order of gravity center position and the order of molecular posture.

Problem 2. Classify the types of liquid crystals and explain them briefly.

Liquid crystals are broadly classified into thermotropic liquid crystals and lyotropic liquid crystals, and are further classified into finer liquid crystalline phases as shown in Table 2. A thermotropic liquid crystal is a substance that exhibits one or more kinds of liquid crystalline phases depending on temperature. Lyotropic liquid crystal is a substance that exhibits one or more kinds of liquid crystalline phases depending on the type of solvent and the concentration.

There are three types of thermotropic liquid crystals: calamitic liquid crystals with rod-like molecules, discotic liquid crystals with disk-like molecules, and flying-seed liquid crystals with bulky substituents instead of long alkyl chains in the periphery. Each liquid crystal substance exhibits one or more types of liquid crystalline phases: the lyotropic liquid crystals exhibit liquid crystalline phases such as a middle phase, a neat phase, and a cubic phase, depending on the type of solvent and the concentration.

There is also a method of classifying liquid crystals according to their molecular weight, and in this classification, liquid crystals are roughly divided into low-molecular-weight liquid crystals and high-molecular-weight liquid crystals.

Problem 3. Using equations of (1) 3D orthogonal system, (2) 3D hexagonal system, (3) 3D tetragonal system, and (4) 3D monoclinic system, explain logically each case of the stepwise disintegration of crystal lattice by heating or solvent addition (See Section 3.3 and Table 3 in the textbook).

Table 3. Relationship between phase dimensionality and X-ray reflection line dimensionality.

Phase dimensionality / Reflection line dimensionality	3D	[2D ⊕ 1D]	2D	1D	0D	[1D ⊕ 1D]	[1D ⊕ 1D ⊕ 1D]
1-dimensional	(00*l*)	(00*l*)	—	(00*l*)	—	(*h*00)+(00*l*)	(*h*00)+(0*k*0)+(00*l*)
2-dimensional	(*hk*0)	(*hk*0)	(*hk*0)	—	—	—	—
3-dimensional	(*hkl*)	—	—	—	—	—	—
Single (S) or composite (C) lattice-based phase	S	C	S	S	S	C	C

(Note) In this table, [2D⊕1D] means the direct sum of a two-dimensional subspace and a one-dimensional subspace, and they are orthogonal to each other. The [2D⊕1D] space is NOT a three-dimensional space. Such a space composed of subspaces using the sign ⊕ is a composite lattice system, so it is written as C at the bottom.

In X-ray diffraction, the three-dimensional $(hk\ell)$ reflections of the crystalline phase are divided into the two-dimensional $(hk0)$ reflections and the one-dimensional (00ℓ) reflections of the mesophase. Table 3 summarizes the dimensionality of X-ray reflection lines and the dimensionality of phases.

To the best of our knowledge to date, all liquid crystalline phases are derived from only four crystal systems: [1] orthorhombic (rectangular), [2] hexagonal, [3] tetragonal and [4] monoclinic. When the crystals are heated or solvent-added, stepwise disintegration of the crystal lattice occurs, as depicted in Figure 3. In the monoclinic crystals [4], one of the angles of the original crystal lattice is tilted instead of 90 degrees. Accordingly, there are three types of monoclinic crystals with either α, β or γ tilted (cf. [4-1], [4-2], [4-3] in Figure 3). As a representative example, only the stepwise disintegration of the monoclinic crystal in [4-2] is shown in this figure.

In the following, we will describe, mathematically and logically, how each of these crystalline systems manifests a liquid crystalline phase by stepwise disintegration.

[1] 3D orthorhombic lattice disintegration

Relationship among the lattice spacing d, the Miller index and the lattice constants in a 3D orthorhombic crystal lattice is described as

$$(3D): \frac{1}{d_{hkl}^2} = \frac{h^2}{a^2} + \frac{k^2}{b^2} + \frac{l^2}{c^2} \tag{1}$$

From this 3D crystal lattice, we can observe all X-ray reflections, *i.e.*, 3D $(hk\ell)$ reflections, 2D $(hk0)$ reflections, and 1D (00ℓ) reflections, as shown in Table 3. However, when heated, the *ab* planes of this lattice begin to slide between each other to collapse the correlation between the *ab* planes, resulting in the disappearance of the 3D long-range three-dimensional order of gravity center position. At this time, the original 3D lattice disintegrates into a 2D lattice and a 1D lattice as shown in Figure 3[1]. In that case, Equation (1) is divided into two

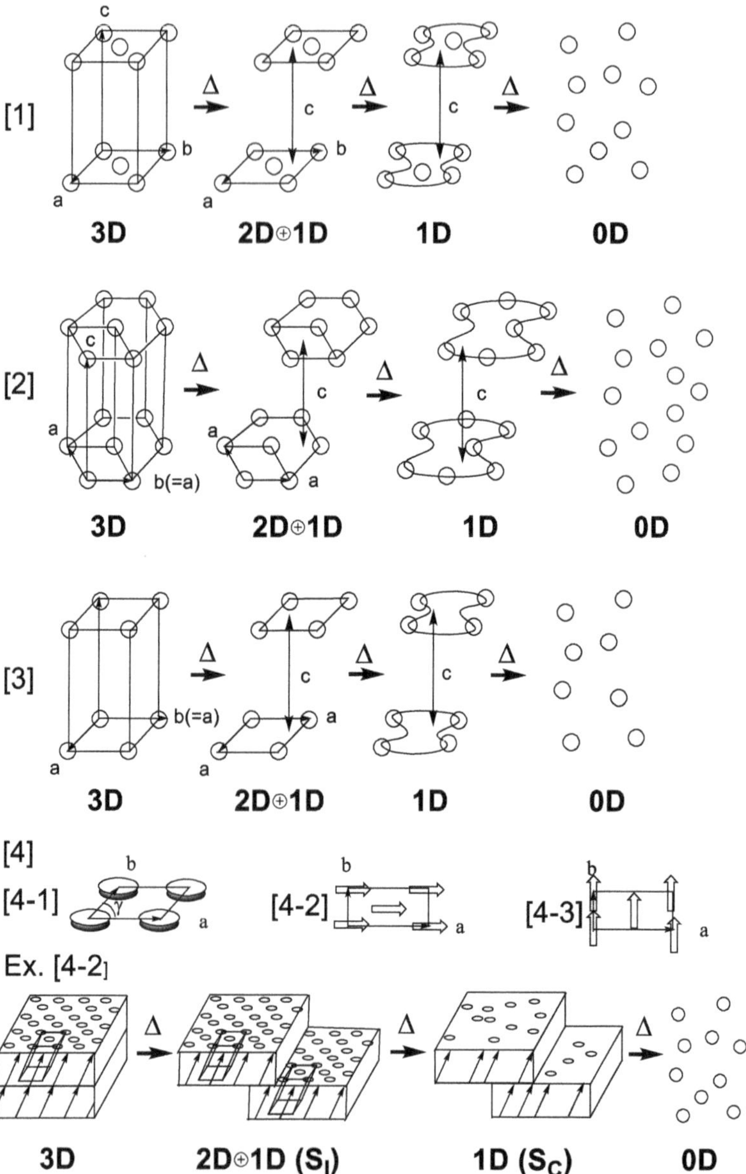

Figure 3. Schematic illustration of stepwise lattice disintegration of a crystal by heating or adding solvent: [1] 3D orthogonal system, [2] 3D hexagonal system, [3] 3D tetragonal system, and [4] 3D monoclinic system. The 3D monoclinic system includes three different types: [4-1] the $\boldsymbol{a}$-axis tilts to the $\boldsymbol{b}$-axis direction, [4-2] the $\boldsymbol{c}$-axis tilts to the $\boldsymbol{a}$-axis direction, and [4-3] the $\boldsymbol{c}$-axis tilts to the $\boldsymbol{b}$-axis direction.

equations (2-1) and (2-2) as follows:

$$\text{(2D)}: \frac{1}{d_{hk0}^2} = \frac{h^2}{a^2} + \frac{k^2}{b^2} \tag{2-1}$$

$$\text{(1D)}: \frac{1}{d_{00l}^2} = \frac{l^2}{c^2} \tag{2-2}$$

At this time, we can observe the 2D ($hk0$) reflections and the 1D (00ℓ) reflections are for the X-ray diffraction, whereas we can no longer observe the 3D ($hk\ell$) reflections. This state corresponds to the S_E, Col_{ro}, and $D_{L.rec}$ phases. On further heating, the 2D lattice distorts to destroy the 2D long-range order of the long-range two-dimensional order of gravity center position. The original 2D lattice thus collapses, leaving only the 1D lattice. In that case, we can observe only 1D (00ℓ) X-ray reflections described in Equation (2-2). This state corresponds to the S_A, S_C, and D_{L1} phases.

At this time, we consider Equation (2-2). Since (ℓ, $c \geq 0$), we can take the square root of both sides to obtain

$$\frac{1}{d_{00l}} = \frac{l}{c} \tag{2-2-1}$$

Since the lattice constant c in this equation is a constant,

$$d_{00l} \propto \frac{1}{l} \tag{2-2-2}$$

Therefore, the spacing $d_{00\ell}$ and the Miller index ℓ are inversely proportional. Here, since $\ell = 1, 2, 3, 4, 5, \ldots$, the spacing ratio is as follows:

$$d_{001} : d_{002} : d_{003} : d_{004} : d_{005} : \cdots = 1 : \frac{1}{2} : \frac{1}{3} : \frac{1}{4} : \frac{1}{5} : \cdots \tag{3}$$

Such a ratio is observed in lamellar liquid crystal phases such as smectic phases and discotic lamellar phases.

On further heating, the 1D lattice described in Equation (2) finally collapses. At this time, we can no longer observe the periodic 1D (00ℓ) reflections like Equation (3), but only the 0D broad reflection is given. This corresponds to the case of nematic phases and isotropic liquids.

[2] 3D hexagonal lattice disintegration

Next, we deal with the disintegration of a 3D hexagonal crystal lattice. The relationship among the lattice spacing d, the Miller index and the lattice constants in a 3D hexagonal crystal lattice is described as

$$(3\mathrm{D}) : \frac{1}{d_{hkl}^2} = \frac{4}{3}\left(\frac{h^2 + hk + k^2}{a^2}\right) + \frac{l^2}{c^2} \tag{4}$$

Similarly in the previous example [1], when the original 3D lattice collapses into a 2D lattice and a 1D lattice upon heating, Equation (4) is divided into two Equations (4-1) and (4-2) as follows:

$$(2\mathrm{D}) : \frac{1}{d_{hk0}^2} = \tfrac{4}{3}\left(\frac{h^2 + hk + k^2}{a^2}\right) \tag{4-1}$$

$$(1\mathrm{D}) : \frac{1}{d_{00l}^2} = \frac{l^2}{c^2} \tag{4-2}$$

From Equation (4-2), we can deduce the same relationship as Equation (3).

The (00ℓ) reflections obtained from this Equation (4-2) correspond to the lamellar structure in an $\mathrm{S_B}$ phase and the stacking structure in the column in the $\mathrm{Col_{ho}}$ phase.

Equation (4-1) is extracted the square root and rearranged as

$$d_{hk0} = \frac{\sqrt{3}a}{2\sqrt{h^2 + hk + k^2}} \tag{4-2-1}$$

Therefore,

$$d_{hk0} \propto \frac{1}{\sqrt{h^2 + hk + k^2}} \tag{4-2-2}$$

When (hk) is substituted with (10), (11), (20), (21), (30), (22), ... into this equation, we can obtain a following ratios:

$$d_{100} : d_{110} : d_{200} : d_{210} : d_{300} : d_{220} : \cdots$$
$$= 1 : \frac{1}{\sqrt{3}} : \frac{1}{2} : \frac{1}{\sqrt{7}} : \frac{1}{3} : \frac{1}{\sqrt{12}} : \cdots \tag{5}$$

Comparing this Equation (5) with the previous Equation (3), the ratios characteristic to the two-dimensional hexagonal (2D-hexagonal) lattice are given $1/\sqrt{3}$, $1/\sqrt{7}$, $1/\sqrt{12}$, *etc.* These ratios correspond to the S_B and Col_h phases, and are a good guideline for identifying the S_B and Col_h phases.

[3] 3D tetragonal lattice disintegration

Furthermore, we consider the disintegration of a 3D tetragonal crystal lattice in the same way mentioned above. Accordingly, the corresponding 2D-tetragonal lattice can be easily deduced as

$$\frac{1}{d_{hk0}^2} = \frac{h^2 + k^2}{a^2} \tag{6}$$

Therefore,

$$d_{hk0} \propto \frac{1}{\sqrt{h^2 + k^2}} \tag{7}$$

From this Equation (7), a following ratios can be obtained:

$$d_{100} : d_{110} : d_{200} : d_{210} : d_{300} : d_{220} : \cdots$$
$$= 1 : \frac{1}{\sqrt{2}} : \frac{1}{2} : \frac{1}{\sqrt{5}} : \frac{1}{3} : \frac{1}{\sqrt{8}} : \cdots \tag{8}$$

Comparing this Equation (8) with Equations (3) and (5), the ratios characteristic to the two-dimensional tetragonal (2D-tetragonal) lattice are given $1/\sqrt{2}$, $1/\sqrt{5}$, $1/\sqrt{8}$, *etc.*

Thus, each of the crystalline 3D lattices disintegrates gradually with increasing temperature, and the lattice dimensionality decreases like as 3D $\rightarrow$ [2D$\oplus$1D] $\rightarrow$ 1D *etc.*, and finally to 0D isotropic liquid. (cf. Figure 3[1]–[3]).

[4] 3D monoclinic lattice disintegration

Finally, we consider the disintegration of a 3D monoclinic crystal lattice. For this lattice, the following three cases must be considered. As can be seen in Figure 3, the monoclinic crystal system has three different tilt directions:

(Case [4-1]) Monoclinic, $a > b >> c$, $\alpha = 90°$, $\beta = 90°$, $\gamma \neq 90°$
The a-axis tilts to the b-axis direction.
(Case [4-2]) Monoclinic, $c >> a > b$, $\alpha = 90°$, $\beta \neq 90°$, $\gamma = 90°$
The c-axis tilts to the a-axis direction.
(Case [4-3]) Monoclinic, $c >> a > b$, $\alpha \neq 90°$, $\beta = 90°$, $\gamma = 90°$
The c-axis tilts to the b-axis direction.

In the stepwise disintegration of these monoclinic crystals, the description by mathematical equations can be considered similar to the above.

When Case [4-1], the monoclinic lattice is expressed as follows:

$$\frac{1}{d_{hkl}^2} = \frac{1}{\sin^2\gamma}\left(\frac{h^2}{\mathrm{a}^2} + \frac{k^2}{\mathrm{b}^2} - \frac{2hk\cos\gamma}{\mathrm{ab}}\right) + \frac{l^2}{\mathrm{c}^2}$$

For this case, when it starts to slide between the ab planes, the three-dimensionality disappears and it is divided into two-dimensional and one-dimensional lattices. Therefore,

$$\frac{1}{\mathrm{d}_{hk0}^2} = \frac{1}{\sin^2\gamma}\left(\frac{h^2}{\mathrm{a}^2} + \frac{k^2}{\mathrm{b}^2} - \frac{2hk\cos\gamma}{\mathrm{ab}}\right)$$

$$\frac{1}{\mathrm{d}_{00l}^2} = \frac{l^2}{\mathrm{c}^2}$$

These equations correspond to a discotic oblique ordered columnar ($\mathrm{Col_{ob.o}}$) liquid crystalline phase.

When Case [4-2], the monoclinic lattice is expressed as follows:

$$\frac{1}{\mathrm{d}_{hkl}^2} = \frac{1}{\sin^2\beta}\left(\frac{h^2}{\mathrm{a}^2} + \frac{l^2}{\mathrm{c}^2} - \frac{2hl\cos\beta}{\mathrm{ac}}\right) + \frac{k^2}{\mathrm{b}^2}$$

For this case, when it starts to slide between the ab planes, the three-dimensionality disappears and it is divided into two-dimensional and one-dimensional lattices.

Therefore, there is no correlation between a and c, so that the contribution of the $\frac{2h\cos\beta}{\mathrm{ac}}$ term becomes zero. As a result, it is divided into a two-dimensional ab lattice and a one-dimensional c lattice, and the equation in this case is expressed as follows:

$$\frac{1}{\mathrm{d}_{hk0}^2} = \frac{h^2}{\mathrm{a}^2} + \frac{k^2}{\mathrm{b}^2}$$

$$\frac{1}{\mathrm{d}_{00l^2}^2} = \frac{l^2}{\mathrm{c}^2}$$

These equations correspond to a smectic F ($\mathrm{S_F}$) phase.

When Case [4-3], the monoclinic lattice is expressed as follows:

$$\frac{1}{\mathrm{d}_{hkl}^2} = \frac{1}{\sin^2\alpha}\left(\frac{h^2}{\mathrm{a}^2} + \frac{l^2}{\mathrm{c}^2} - \frac{2hl\cos\alpha}{\mathrm{bc}}\right) + \frac{k^2}{\mathrm{a}^2}$$

For this case, when it starts to slide between the ab planes, the three-dimensionality disappears and it is divided into two-dimensional and one-dimensional lattices.

Therefore, there is no correlation between b and c, so that the contribution of the $\frac{2h\cos\alpha}{\mathrm{bc}}$ term becomes zero. As a result, it is divided into a two-dimensional ab lattice and a one-dimensional c lattice, and the equation in this case is expressed as follows:

$$\frac{1}{\mathrm{d}_{hk0}^2} = \frac{h^2}{\mathrm{a}^2} + \frac{k^2}{\mathrm{b}^2}$$

$$\frac{1}{\mathrm{d}_{00l}^2} = \frac{l^2}{\mathrm{c}^2}$$

These equations correspond to a smectic I ($\mathrm{S_I}$) phase.

As mentioned above, in both of Case [4-2] and Case [4-3], two-dimensional and one-dimensional equations are exactly the same. However, the X-ray photographic method using the monodomain enables us to determine whether the c-axis is tilted in the a-axis direction or the b-axis direction. Thus, we can identify between the $\mathrm{S_F}$ phase and the $\mathrm{S_I}$ phase. (cf. P. A. C. Gane, A. J. Leadbetter and P. G. Wrighton, *Mol. Cryst., Liq. Cryst.*, **66**, 247–266 (1981)).

Table 4. Relationship among spacing (d), Miller indices (hkl) and the lattice constants (a, b, c, α, β and γ) for [2D$\oplus$1D] mesophases known to date.

3D	2D $\oplus$ 1D	Lattices after degradation	Mesophase
Orthorhombic $\frac{1}{d_{hkl}^2}=\frac{h^2}{a^2}+\frac{k^2}{b^2}+\frac{l^2}{c^2}$ ⇨	$\frac{1}{d_{hk0}^2}=\frac{h^2}{a^2}+\frac{k^2}{b^2}$ $\frac{1}{d_{00l}^2}=\frac{l^2}{c^2}$	Rectangular Stacking distance(c = h), Layer thickness (c)	**Col_{ro} (a>b>>c), S_E (c>>a>b)**
3D-hexagonal $\frac{1}{d_{hkl}^2}=\frac{4}{3}\left(\frac{h^2+hk+k^2}{a^2}\right)+\frac{l^2}{c^2}$ ⇨	$\frac{1}{d_{hk0}^2}=\frac{4}{3}\left(\frac{h^2+hk+k^2}{a^2}\right)$ $\frac{1}{d_{00l}^2}=\frac{l^2}{c^2}$	2D-hexagonal Stacking distance (c = h), Layer thickness (c)	**Col_{ho} (a>>c), S_B (c>>a) (S_L)**
3D-tetragonal $\frac{1}{d_{hkl}^2}=\frac{h^2+k^2}{a^2}+\frac{l^2}{c^2}$ ⇨	$\frac{1}{d_{hk0}^2}=\frac{h^2+k^2}{a^2}$ $\frac{1}{d_{00l}^2}=\frac{l^2}{c^2}$	2D-tetragonal Stacking distance (c = h), Layer thickness (c)	**$Col_{tet.o}$ (a>>c), S_T (c>>a)**
Monoclinic **$\alpha = 90°$, $\beta = 90°$, $\gamma \neq 90°$** **a>b>>c** The a axis tilts to the b axis direction. $\frac{1}{d_{hkl}^2}=\frac{1}{\sin^2\gamma}\left(\frac{h^2}{a^2}+\frac{k^2}{b^2}-\frac{2hk\cos\gamma}{ab}\right)+\frac{l^2}{c^2}$ ⇨	$\frac{1}{d_{hk0}^2}=\frac{1}{\sin^2\gamma}\left(\frac{h^2}{a^2}+\frac{k^2}{b^2}-\frac{2hk\cos\gamma}{ab}\right)$ $\frac{1}{d_{00l}^2}=\frac{l^2}{c^2}$ 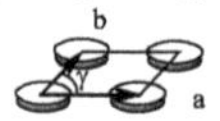 b a (a>b>>c) $\gamma \neq 90°$	Oblique Stacking distance (c = h)	**$Col_{ob.o}$**
Monoclinic **$\alpha = 90°$, $\beta \neq 90°$, $\gamma = 90°$** **c>>a>b** The c axis tilts to the a axis direction. $\frac{1}{d_{hkl}^2}=\frac{1}{\sin^2\beta}\left(\frac{h^2}{a^2}+\frac{l^2}{c^2}-\frac{2hl\cos\beta}{ac}\right)+\frac{k^2}{b^2}$ ⇨	$\frac{1}{d_{hk0}^2}=\frac{h^2}{a^2}+\frac{k^2}{b^2}$ $\frac{1}{d_{00l}^2}=\frac{l^2}{c^2}$ 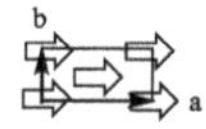 b a (c>>a>b) $\beta \neq 90°$	Rectangular Layer thickness (c)	**S_F (S_G) (S_H)**
Monoclinic **$\alpha \neq 90°$, $\beta = 90°$, $\gamma = 90°$** **c>>a>b** The c axis tilts to the b axis direction. $\frac{1}{d_{hkl}^2}=\frac{1}{\sin^2\alpha}\left(\frac{k^2}{b^2}+\frac{l^2}{c^2}-\frac{2kl\cos\alpha}{bc}\right)+\frac{h^2}{a^2}$ ⇨	$\frac{1}{d_{hk0}^2}=\frac{h^2}{a^2}+\frac{k^2}{b^2}$ $\frac{1}{d_{00l}^2}=\frac{l^2}{c^2}$ 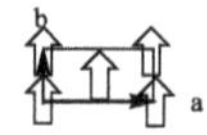 b a (c>>a>b) $\alpha \neq 90°$	Rectangular Layer thickness (c)	**S_I (S_J) (S_K)**

From the above, we can recognize that $Col_{ob.o}$, S_F and S_I liquid crystalline phases are originated from the stepwise disintegration of the monoclinic three-dimensional crystal lattices. Figure 3[4] depicts the stepwise disintegration of monoclinic lattice in Case [4-2], as a representative example.

Finally, Table 4 summarizes the liquid crystal phases appearing by the stepwise disintegration of each of the four crystal systems: [1] orthorhombic, [2] hexagonal, [3] tetragonal, and [4] monoclinic, together with their two-dimensional and one-dimensional lattice equations. In addition, the S_L, S_G, S_H, S_J, and S_K phases in the

Table 5. Original crystal systems and dimensionalities of liquid crystalline phases resulted by stepwise lattice disintegration.

Rod-like LC mesophase	Tilt	Original crystal system	Order in the layer	Phase Dimensionality$	Discotic LC mesophase	Tilt	Original crystal system	Order in the face	Phase Dimensionality$
N				0D	N_D				0D
S_A	⊥		Non (1D-layer)	1D					
S_C	∠		Non (1D-layer)	1D	D_{L1}	∠		Non (1D-layer)	1D
					Col_{hd}		3D-hexagonal	2D-hexagonal	2D
					$Col_{tet\,d}$		3D-tetragonal	2D-tetragonal	2D
					Col_{rd}		3D-orthorohmbic	Rectangular	2D
					$Col_{ob\,d}$	$\angle_\gamma$	3D-monoclinic	2D-pallarelogram	2D
S_D	---	3D-cubic	---	3D					
S_B	⊥	3D-hexagonal	2D-hexagonal	[2D⊕1D]	Col_{ho}		3D-hexagonal	2D-hexagonal	[2D⊕1D]
S_T	⊥	3D-tetragonal	2D-tetragonal	[2D⊕1D]	$Col_{tet\,o}$		3D-tetragonal	2D-tetragonal	[2D⊕1D]
S_E	⊥	3D-orthorohmbic	Rectangular	[2D⊕1D]	Col_{ro}		3D-orthorohmbic	Rectangular	[2D⊕1D]
					$D_{L\,rec}(2_1 2_1)$	⊥	3D-orthorohmbic	Rectangular	[2D⊕1D]
					$D_{L\,rec}(12_1)$	⊥	3D-orthorohmbic	Rectangular	[2D⊕1D]
S_F	$\angle_\beta$	3D-monoclinic.	Rectangular	[2D⊕1D]	$Col_{ob\,o}$	$\angle_\gamma$	3D-monoclinic	2D-pallarelogram	[2D⊕1D]
S_I	$\angle_\alpha$	3D-monoclinic.	Rectangular	[2D⊕1D]					
					D_{L2} (= Col_L)	⊥			[1D⊕1D]
					No name	⊥			[1D⊕1D⊕1D]
S_L	⊥	3D-hexagonal	2D-hexagonal	3D < [2D⊕1D]					
S_G	$\angle_\beta$	3D-monoclinic.	Rectangular	3D < [2D⊕1D]					
S_H	$\angle_\beta$	3D-monoclinic	Rectangular	3D < [2D⊕1D]					
S_J ($_{G'}$)	$\angle_\alpha$	3D-monoclinic.	Rectangular	3D < [2D⊕1D]					
S_K ($_{H'}$)	$\angle_\alpha$	3D-monoclinic	Rectangular	3D < [2D⊕1D]					

$: Phase dimensionality can be determined by the X-ray reflection lines: 1D: (00l); 2D: (hk0); 3D: (hkl). α: angle α is not 90°; β: angle β is not 90°; γ: angle γ is not 90°.

table are "hard liquid crystal phases" in the gray zone, where they give the X-ray reflections with a little three-dimensionality; there has been a long debate as to whether they are liquid crystals or crystals (See Table I in "G. W. Gray and J. W. Goodby, *Smectic Liquid Crystals Texture and Structures*, Leonard Hill, Grasgow (1981)"; See the discussion of Section 3.6.1 in the textbook).

Problem 4. Give several examples of the liquid crystalline compounds showing the 1D, 2D, [2D⊕1D], [1D⊕1D] and [1D⊕1D⊕1D] mesophases. Be sure to draw each of the molecular structures and indicate each of the source original papers.

Table 5 summarizes the original crystal systems and dimensionalities of liquid crystalline phases resulted by stepwise lattice disintegration.

As can be seen from this table, the liquid crystalline phases in each dimensionality are as follows:

- 1D: S_A, S_C, D_{L1}
- [1D⊕1D]: D_{L2} (= Col_L)
- 2D: Col_{hd}, $Col_{tet.d}$, Col_{rd}, $Col_{ob.d}$
- [2D⊕1D]: S_B, S_T, S_E, S_F, S_I; Col_{ho}, $Col_{tet.o}$, Col_{ro}, $D_{L.rec}(2_12_1)$, $D_{L.rec}(12_1)$
- [1D⊕1D⊕1D]: No name (Tentatively named as **Trisoned** mesophase)

The molecular structures of the liquid crystalline compounds are shown below together with the corresponding literatures. Each has been selected that it described the X-ray liquid crystal structure analysis.

- 1D mesophases:
  S_A,

  [1] P. Davidson, A. M. Levelut, M. F. Achard and F. Hardouin, *Liq. Cryst.*, **4**, 561–571 (1989).

  $(CH_3)_3$-Si-[-O-Si(CH_3)((CH$_2$)$_n$-O-C$_6$H$_4$-COO-C$_6$H$_4$-OC$_m$H$_{2m+1}$)-]$_{35}$-O-Si-$(CH_3)_3$ $P_{n,m}$: $P_{3,4}$, $P_{4,4}$, $P_{3,8}$, $P_{5,8}$

  [2] M. Ghedini, D. Pucci, R. Bartolino and O. Francescangeli, *Liq. Cryst.*, **13**, 255–263 (1993).

  R, OR', N, N, Pd, X, X, Pd, N, N, OR', R

  $\begin{bmatrix} R \\ R' \end{bmatrix} = \begin{bmatrix} C_9H_{19} \\ CH_3 \end{bmatrix}$ (HL_1), $\begin{bmatrix} C_6H_{13} \\ C_{11}H_{23} \end{bmatrix}$ (HL_2), $\begin{bmatrix} C_9H_{19} \\ C_9H_{19} \end{bmatrix}$ (HL_3)

  X = Cl, Br, I

  S_C,

  [3] G. Peul, P. Kolbe, U. Preukschas, S. Diele and D. Demus, *Mol. Cryst. Liq. Cryst.*, **53**, 167–180 (1979).

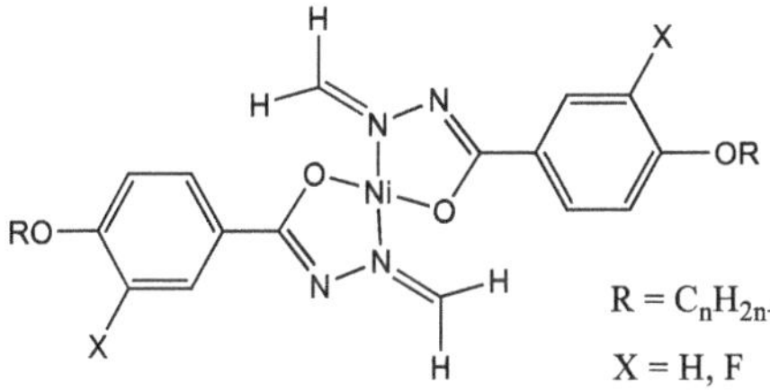

$R = C_nH_{2n+1}$ (n = 14, 22)

X = H, F

[4] M. N. Abser, M. Bellwood, C. Buckley, M. C. Holmes and R. W. McCabe, *Mol. Cryst. Liq. Cryst*, **260**, 333–337 (1995).

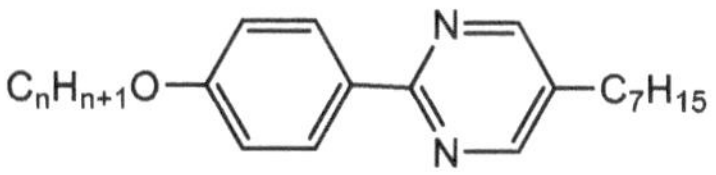

$C_nH_{n+1}O$ — C_7H_{15} n = 9, 10

D_{L1},

[5] K. Ohta, H. Muroki, A. Takagi, K. Hatada, H. Ema, I. Yamamoto and K. Matsuzaki, *Mol. Cryst. Liq. Cryst.*, **140**, 131–152 (1986).

[6] H. Sakashita, A. Nishitani, Y. Sumiya, H. Terauchi, K. Ohta and I. Yamamoto, *Mol. Cryst. Liq. Cryst.*, **163**, 211–219 (1988).

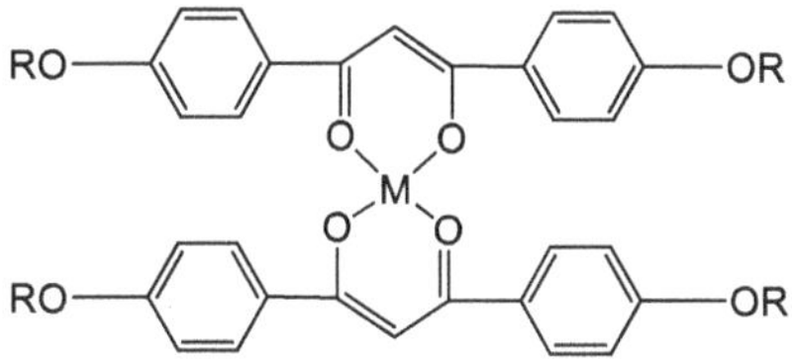

$R = C_nH_{2n+1}$ (n = 8~12); M = Cu

- [1D⊕1D] mesophases:
  D_{L2} (= Col_L),

[7] J. Billard, *C. R. Acad. Sc. Paris*, t. 299, Serie II, no 14, 905 (1984).

[8] K. Ohta, A. Ishii, I. Yamamoto and K. Matsuzaki, *J. Chem. Soc., Chem. Commun.*, 1099 (1984).

[9] K. Ohta, H. Muroki, A. Takagi, K. Hatada, H. Ema, I. Yamamoto and K. Matsuzaki, *Mol. Cryst. Liq. Cryst.*, **140**, 131–152 (1986).

[10] H. Sakashita, A. Nishitani, Y. Sumiya, H. Terauchi, K. Ohta and I. Yamamoto, *Mol. Cryst. Liq. Cryst.*, **163**, 211–219 (1988).

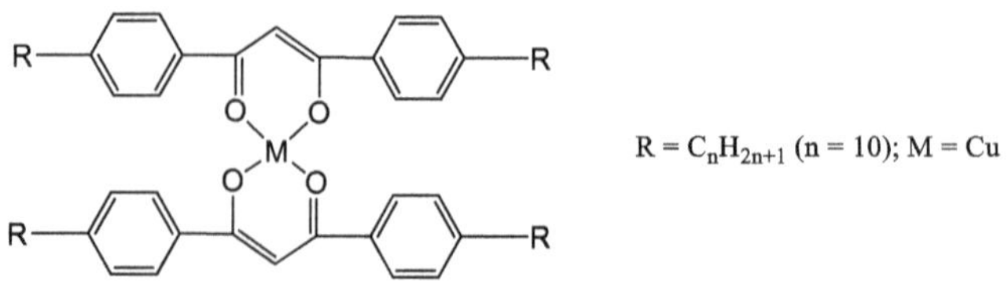

R = C_nH_{2n+1} (n = 10); M = Cu

[11] T. Nakai, K. Ban, K. Ohta and M. Kimura, *J. Mater. Chem.*, **12**, 844–850 (2002).

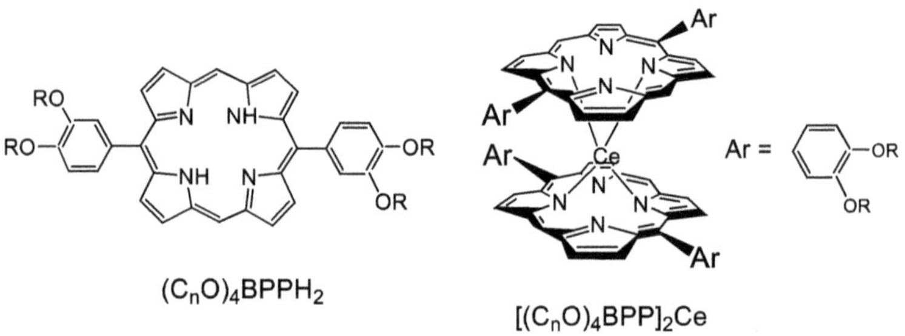

$(C_nO)_4BPPH_2$

$[(C_nO)_4BPP]_2Ce$

[12] S. Méry, D. Haristoy, J.-F. Nicoud, D. Guillon, S. Diele, H. Monobe and Y. Shimizu, *J. Mater. Chem.*, **12**, 37.

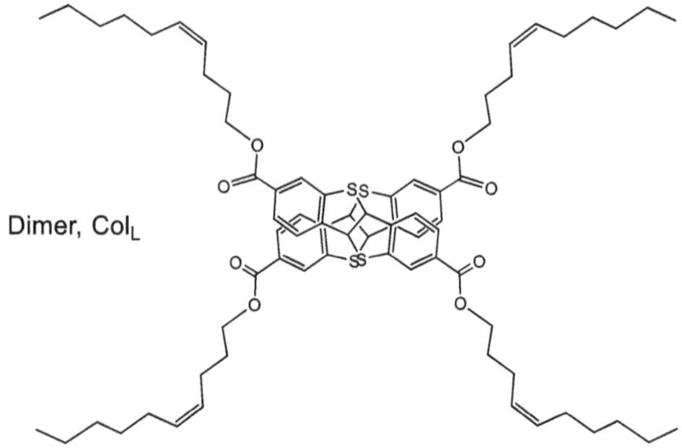

Dimer, Col_L

This liquid crystalline phase has both lamellar and columnar structures, so that it is named as a discotic lamellar (D_L) phase [7–11] or a lamellocolumnar (Col_L) phase [12]. In this phase, the layers slip between each other and the columns also slip between each other. However, the distance between the columns in b axis direction is not constant but random.

This phase was first discovered in the bis(β-diketonato) copper(II) complexes in the 1980s [7–10], but the mesomorphism had been long controversial till the 2000s. Many people strongly believed in the 1980s that discotic liquid crystals should need more than six long chains in the periphery. Accordingly, there was a radical argument that the four long chain-substituted copper(II) complexes [7–11] were not liquid crystalline but crystalline. However, by the 2000's this liquid crystalline phase had been discovered in the other compounds [11, 12]. Since then, this liquid crystalline phase, D_L and Col_L, has been world-widely accepted.

Interestingly, this [1D⊕1D] liquid crystalline phase tends to appear in a disk-like compound having four long chains in the periphery. Although the Méry's compound has only two long chains in the periphery, it forms dimer to have four long chains in the periphery to show the Col_L liquid crystalline phase [12].

- 2D mesophases:
  Col_{hd},

[13] K. Ohta, M. Moriya, M. Ikejima, H. Hasebe, T. Fujimoto and I. Yamamoto, *Bull. Chem. Soc., Jpn.*, **66**, 3559–3564 (1993).

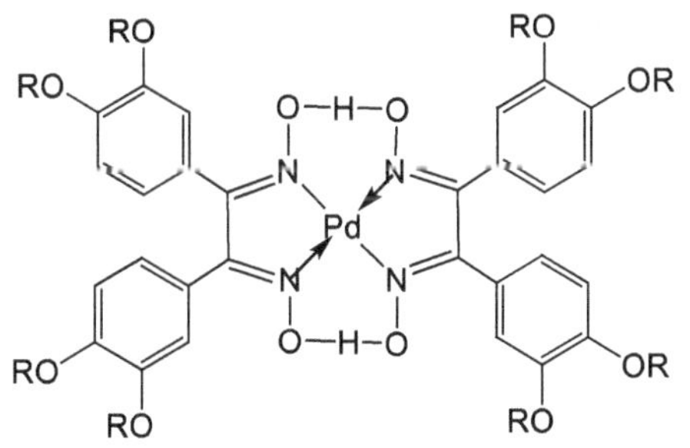

R = C_nH_{2n+1} (n = 2-12)

$Col_{tet.d}$, Col_{rd},

[14] K. Hatsusaka, K. Ohta, I. Yamamoto and H. Shirai, *J. Mater. Chem.*, **11**, 423–433 (2001).

$[(C_nO)_2PhO]_8PcCu$

$R = C_nH_{2n+1}$ (n = 9~14)

$Col_{ob.d}$,

[15] T. Komatsu, K. Ohta, T. Fujimoto and I. Yamamoto, *J. Mater. Chem.*, **4**, 533–536 (1994).

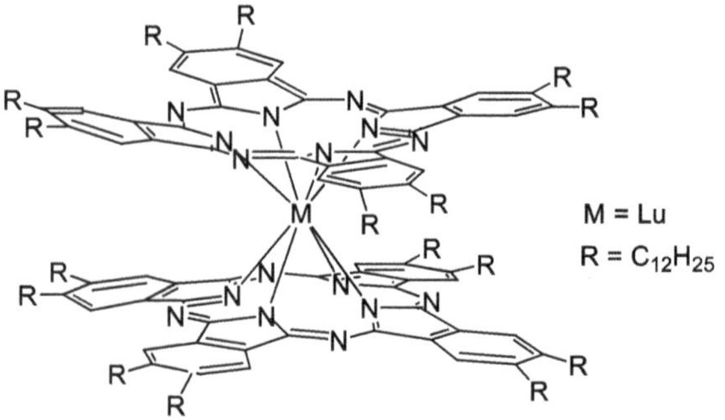

- [2D⊕1D] mesophases:
  S_B,

[16] W. H. De Jeu and J. A. De Pooter, *Phys. Lett.*, **61A**, 114–116 (1977).

[17] W. Klämke and W. Haase, *Z. Naturforsch.* **43a**, 885–888 (1988).

S_T,

[18] E. Alami, H. Levy, R. Zana, P. Weber and A. Soulios, *Liq. Cryst.*, **13**, 201–212 (1993).

x = 12, 14, 16, 18;
y = 12, 14, 16, 18

[19] K. Ohta, T. Sugiyama and T. Nogami, *J. Mater. Chem.*, **10**, 613–616 (2000).

n = 10, 12, 14, 16, 18

[20] K. Sugiyama, M. Yasutake and K. Ohta, *Liq. Cryst.*, (2022). https://doi.org/10.1080/02678292.2022.2084167

This ionic compound was reported to show two crystalline phases by Iwamato *et al.*, in 1981, but Ohta *et al.*, reinvestigated to establish the higher temperature phase is not crystalline phase but a smectic 'T' phase in 2022. It is the phantom first example of a smectic T phase.

S_E,

[21] K. Ohta, O. Takenaka, H. Hasebe, Y. Morizumi, T. Fujimoto and I. Yamamoto, *Mol. Cryst. Liq. Cryst.*, **195**, 103–121 (1991).

R = C_nH_{2n+1} (n = 8~12, 16, 18)

[22] K. Ohta, Y. Morizumi, T. Fujimoto and I. Yamamoto, *Mol. Cryst. Liq. Cryst.*, **214**, 161–169 (1992).

C_nH_{n+1} C_nH_{n+1} n = 6, 12 M = Ni

S_F,

[23] J. J. Benattar, J. Doucet, M. Lambert and A. M. Levelut, *Phys. Rev. A,* **20**, 2505–2509 (1979).

[24] F. Moussa, J. J. Benattar and C. Williams, *Mol. Cryst. Liq. Cryst.,* **99**, 145–154 (1983).

C_nH_{2n+1} C_nH_{2n+1}

n = 5, 10

S_I,

[25] P. A. C. Gane, A. J. Leadbetter, J. J. Benattar, F. Moussa and M. Lambert, *Phys. Rev. A,* **24**, 2694–2700 (1981).

C_nH_{2n+1} C_nH_{2n+1}

n = 10

[26] S. Gierlotka, J. Przedmojski and B. Pura, *Liq. Cryst.*, **3**, 1535–1541 (1988).

$C_8H_{17}O$ COO $CH_2CH(CH_3)C_2H_5$

Col_{ho},

[27] M. Ichihara, H. Suzuki, B. Mohr and K. Ohta, *Liq. Cryst.*, **34**, 401–410 (2007).

$R=C_nH_{2n+1}$
n= 8, 10, 12, 14

$(C_nO)_4DCT$ $(C_nO)_4DADCT$

$Col_{tet.o}$,

[28] K. Hatsusaka, M. Kimura and K. Ohta, *Bull. Chem. Soc., Jpn.*, **76**, 781–787 (2003).

R= $-C_6H_3(OC_nH_{2n+1})_2$

n=12, 13

$\{[(C_nO)_2PhO]_8Pc\}_2Lu$

Col_{ro},

[29] K. Ohta, O. Takenaka, H. Hasebe, Y. Morizumi, T. Fujimoto, and I. Yamamoto, *Mol. Cryst. Liq. Cryst.*, **195**, 135–148 (1991).

M = Cu

R = C_nH_{2n+1} (n = 8~12, 16, 18)

$D_{L.rec}(2_11)$ and $D_{L.rec}(2_12_1)$,

[30] K. Ohta, R. Higashi, M. Ikejima, I. Yamamoto and N. Kobayashi, *J. Mater. Chem.*, **8**, 1979–1991 (1998).

$D_{L.rec}$.($P2_11$) | $D_{L.rec}$.($P2_12_1$)

R=C_nH_{2n+1} , n = 12, 16 | R=C_nH_{2n+1} , n =10, 12, 14

4 : Ni(n, 0) | 5 : Ni(n)

- [1D⊕1D⊕1D] mesophases:
  No name (this mesophase is tentatively named here as **Trisoned** [∵ Tris-One-D = 1D⊕1D⊕1D])

[31] D. Pressner, Chr. Göltner, H. W. Spieß and K. Müllen, *Ber. Bunsenges. Phys. Chem.*, **97**, 1362: Compound **P0-17** (1993).

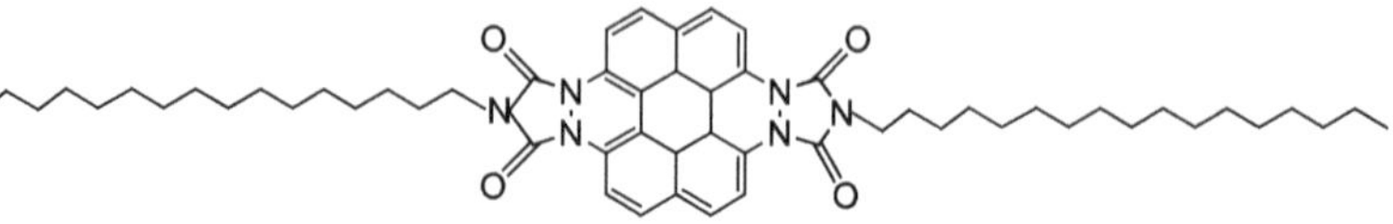

In 1993, Spieß and his co-workers reported an X-ray diffraction pattern of the mesophase in the compound **P0-17** [31]. As can be seen from this X-ray diffraction pattern, this mesophase shows (100), (200), (300), (010) and (001) reflections. Therefore, for the mesophase are observed only (h00), (0k0) and (00l) reflections, which are apparently resulted from a [1D⊕1D⊕1D] **composite-lattice-based** state. Although this mesophase was erroneously assigned as a smectic phase at that time, it is the first example of the [1D⊕1D⊕1D] **composite-lattice-based** liquid crystalline phase, to our best knowledge.

[32] A. El-ghayoury, L. Douce, A. Skoulios and R. Ziessel, *Angew. Chem. Int. Ed.*, **37**, 1255 (1998).

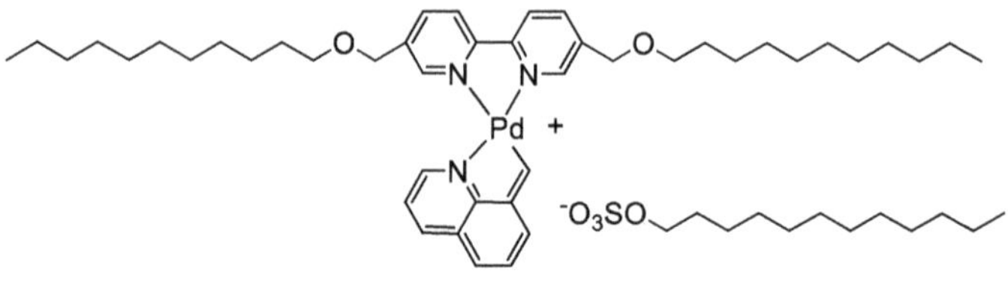

In 1998, Ziessel and his co-workers found a very unique liquid crystalline phase in ortho-palladated bipyridine complexes, and depicted the liquid crystalline phase structure model at the first time [32]. However, it is the same liquid crystalline phase as that of the compound P0-17, although they did not notice it at that time. The present book points out at the first time that both liquid crystalline phases can be classified to the same [1D⊕1D⊕1D] **composite-lattice-based** liquid crystalline phase. This liquid crystalline phase has not yet been named. Therefore, the author tentatively names this phase [1D⊕1D⊕1D] mesophase as **Trisoned** [Tris-One-D] mesophase.

Problem 5. Calculate the extinction rules of a two-dimensional rectangular lattice for (1) $C2/m$, (2) $P2_1/a$, (3) $P2/b$ symmetry, from these liquid crystal structure factors.

The relationship between the crystal structure factor and the Miller index is described as

$$f(hkl) = \sum_{i}^{N} f_i \cdot exp\{2\pi i(hx_i + ky_i + lz_i)\} \quad (10)$$

(**Crystal structure factor**)

Reducing this 3D expression to the 2D expression,

$$f(hk) = \sum_{i}^{N} f_i \cdot exp\{2\pi i(hx_i + ky_i)\} \tag{10'}$$

(**Liquid crystal structure factor**)

From this equation $10'$, two-dimensional extinction rules can be deduced.

(1) In the case of a $C2/m$ symmetry

A two-dimensional rectangular lattice with the symmetry $C2/m$ has the structure shown in Figure 4.

For this symmetry, we make a table below to calculate liquid crystalline structure factor from Equation ($10'$).

Each of the h and k is integer. Also, from Euler's formula, $\exp(i\theta) = \cos\theta + i\sin\theta$, so

$$\exp(2\pi ih) = \exp(2\pi ik) = 1$$

$$\exp(\pi ih) = \cos(h\pi) = (-1)^h$$

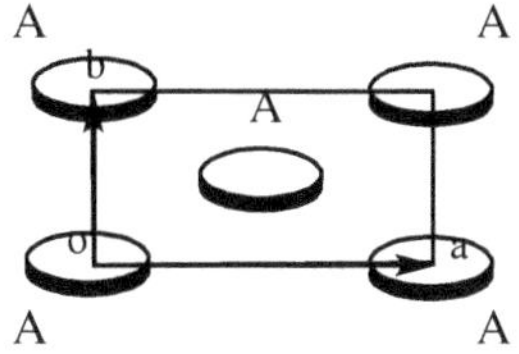

Figure 4. A 2D rectangular lattice having $C2/m$ symmetry.

Table 6. Derivation of the extinction rule for symmetry of $C2/m$.

Coordinate of A	n in nf_A	$\exp\{2\pi\mathrm{i}(\mathrm{h}x_i + ky_i)\}$
(0 0)	$\frac{1}{4}$	$\exp\{0\} = 1$
(1 0)	$\frac{1}{4}$	$\exp\{2\pi\mathrm{i}(\mathrm{h})\} = 1$
(1 1)	$\frac{1}{4}$	$\exp\{2\pi\mathrm{i}(\mathrm{h} + k)\} = 1$
(0 1)	$\frac{1}{4}$	$\exp\{2\pi\mathrm{i}(k)\} = 1$
$\left(\frac{1}{2}\frac{1}{2}\right)$	1	$\exp\left\{2\pi\mathrm{i}\left(\frac{1}{2}\mathrm{h} + \frac{1}{2}k\right)\right\} = (-1)^{\mathrm{h}+k}$

Therefore, from this table and Equation 10′,

$$\mathrm{F(hk)} = f_A\left[\frac{1}{4} \times 1 + \frac{1}{4} \times 1 + \frac{1}{4} \times 1 + \frac{1}{4} \times 1 + 1 \times (-1)^{h+k}\right]$$
$$= f_A[1 + (-1)^{h+k}]$$

Accordingly, when $\mathrm{h} + \mathrm{k} = 2n$, $\mathrm{F(hk)} = 2f$
when $\mathrm{h} + \mathrm{k} = 2n + 1$, $\mathrm{F(hk)} = 0$ (extinction)
Hence, the extinction rule can be derived as

$$\mathbf{h} + \mathbf{k} = 2\boldsymbol{n} + 1 \tag{11}$$

(2) In the case of a $P2_1/a$ symmetry

A two-dimensional rectangular lattice with the $P2_1/a$ symmetry has the structure shown in Figure 5.

For this symmetry, we make a table below to calculate liquid crystalline structure factor from Equation (10′).

By using Equation (10′) and Table 7, we can calculate the liquid crystalline structure factor as follows:

$$\mathrm{F(hk)} = f_A\left(\frac{1}{4} \times 1 + \frac{1}{4} \times 1 + \frac{1}{4} \times 1 + \frac{1}{4} \times 1\right)$$
$$+ f_B[1 \times (-1)^{h+k}] = f_A + f_B(-1)^{h+k} \tag{12}$$

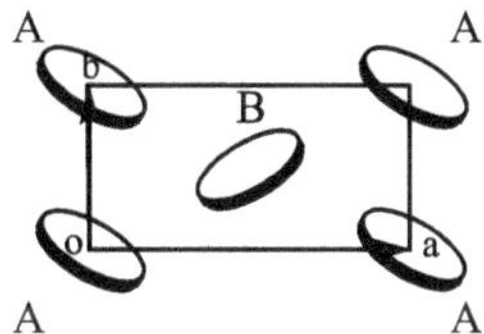

Figure 5. A 2D rectangular lattice having $P2_1/a$ symmetry.

Table 7. Derivation of the extinction rule for symmetry of $P2_1/a$.

Coordinate of A	n in $nf_{\mathbf{A}}$	$\exp\{2\pi i(hx_i + ky_i)\}$
(0 0)	$\frac{1}{4}$	$\exp\{0\} = 1$
(1 0)	$\frac{1}{4}$	$\exp\{2\pi i(h)\} = 1$
(1 1)	$\frac{1}{4}$	$\exp\{2\pi i(h + k)\} = 1$
(0 1)	$\frac{1}{4}$	$\exp\{2\pi i(k)\} = 1$
Coordinate of B	n in nf_{B}	$\exp\{2\pi i(hx_i + ky_i)\}$
$\left(\frac{1}{2}\frac{1}{2}\right)$	1	$\exp\left\{2\pi i\left(\frac{1}{2}h + \frac{1}{2}k\right)\right\} = (-1)^{h+k}$

Accordingly,

when $h + k = 2n$,

$$F(hk) = f_A + f_B \text{ (appears in high intensity).}$$

when $h + k = 2n + 1, F(hk) - f_A - f_B$ (appears in low intensity). (13)

Looking at these Equations (12) and (13), it seems that there is no condition for $F(hk) = 0$. However, considering this symmetry, we can find $f_A = f_B$ for the special conditions that is the case when $h = 0$ or $k = 0$.

As shown in Figure 6, for example, when $h = 0$, the A molecule on the primary reflection surface (01) tilts by $-\beta$ with respect to the b-axis, and the B molecule on the secondary reflection surface (02) tilts by $+\beta$ with respect to the b-axis. In this case, $f_A = f_B$. Even when $k = 0$, the A molecule on the primary reflection surface

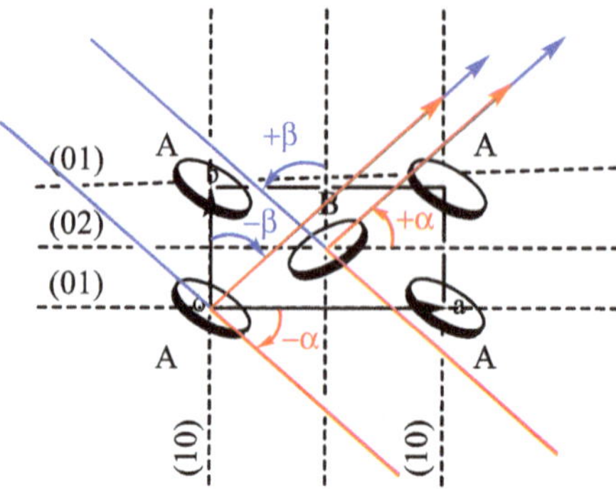

Figure 6. In a 2D rectangular lattice having $P2_1/a$ symmetry, the reflectivity at h = 0 or k = 0 is the same for both molecules A and B ($f_A = f_B$).

(01) tilts by $-\alpha$ with respect to the a-axis, and the B molecule on the secondary reflection surface (02) tilts by $+\alpha$ with respect to the a-axis. Also in this case, $f_A = f_B$.

Therefore, from Equation (13)

When h0: h = 2n + 1, F(hk) = 0 (extinction)

When 0k: k = 2n + 1, F(hk) = 0 (extinction)

Therefore, the extinction rule for the $P2_1/a$ symmetry can be derived as

$$\mathbf{h0: h = 2n + 1}$$

$$\mathbf{0k: k = 2n + 1}$$

(3) In the case of a $P2/a$ symmetry

A two-dimensional rectangular lattice with this symmetry has the structure shown in Figure 7.

Conventionally, the a-axis and the b-axis have been set with being $a < b$ as shown in Figure 4(4). Hereupon, we will consider the conventional case at first. In this case, from Table 8 and Equation (10′), we can calculate the liquid crystal structure factor F(hk).

Accordingly, the sum of F(hk) is

$$\mathrm{F(hk)} = f_A + f_B \cdot (-1)^k + f_C \cdot (-1)^h \left\{ i\sin\left(\frac{\pi}{2}k\right) + i\sin\left(\frac{3\pi}{2}k\right) \right\}$$

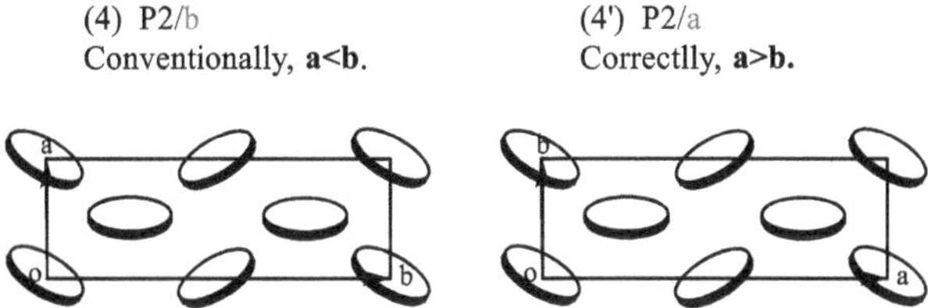

Figure 7. 2D rectangular lattices having $P2/b$ and $P2/a$ symmetries.

Table 8. Derivation of the extinction rule for symmetry of $P2/b$.

Coordinate of A	n in $nf_{\mathbf{A}}$	$\exp\{2\pi\mathrm{i}(\mathrm{h}x_i + ky_i)\}$
(0 0)	$\frac{1}{4}$	$\exp\{0\} = 1$
(1 0)	$\frac{1}{4}$	$\exp\{2\pi\mathrm{i}(\mathrm{h})\} = 1$
(1 1)	$\frac{1}{4}$	$\exp\{2\pi\mathrm{i}(\mathrm{h} + k)\} = 1$
(0 1)	$\frac{1}{4}$	$\exp\{2\pi\mathrm{i}(k)\} = 1$
Coordinate of B	n in nf_{B}	$\exp\{2\pi\mathrm{i}(\mathrm{h}x_i + ky_i)\}$
$(0\frac{1}{2})$	$\frac{1}{2}$	$\exp\left\{2\pi\mathrm{i}\left(\frac{1}{2}k\right)\right\} = (-1)^k$
$(1\frac{1}{2})$	$\frac{1}{2}$	$\exp\left\{2\pi\mathrm{i}\left(\mathrm{h} + \frac{1}{2}k\right)\right\} = (-1)^k$
Coordinate of C	n in nf_{C}	$\exp\{2\pi\mathrm{i}(\mathrm{h}x_i + ky_i)\}$
$(\frac{1}{2}\frac{1}{4})$	1	$\exp\left\{2\pi\mathrm{i}\left(\frac{1}{2}\mathrm{h} + \frac{1}{4}k\right)\right\} = (-1)^{\mathrm{h}} \cdot i\sin\left(\frac{\pi}{2}k\right)$
$(\frac{1}{2}\frac{3}{4})$	1	$\exp\left\{2\pi\mathrm{i}\left(\frac{1}{2}\mathrm{h} + \frac{3}{4}k\right)\right\} = (-1)^{\mathrm{h}} \cdot i\sin\left(\frac{3\pi}{2}k\right)$

As shown in Figure 8 (4), the A molecule on the first reflection surface (01) tilts by $-\alpha$ with respect to the b axis, and the B molecule on the secondary reflection surface (02) tilts by $+\alpha$ with respect to the b axis.

Therefore, when $\mathrm{h} = 0$, $f_A = f_B = f$.

$$\therefore \mathrm{F(hk)} = f\{1 + (-1)^k\} + f_C \cdot \left\{i\sin\left(\frac{\pi}{2}k\right) + i\sin\left(\frac{3\pi}{2}k\right)\right\}$$

When $\mathrm{h} = 0$ and $\mathrm{k} = 2n + 1$, $\mathrm{F(hk)} = 0$ (extinction).

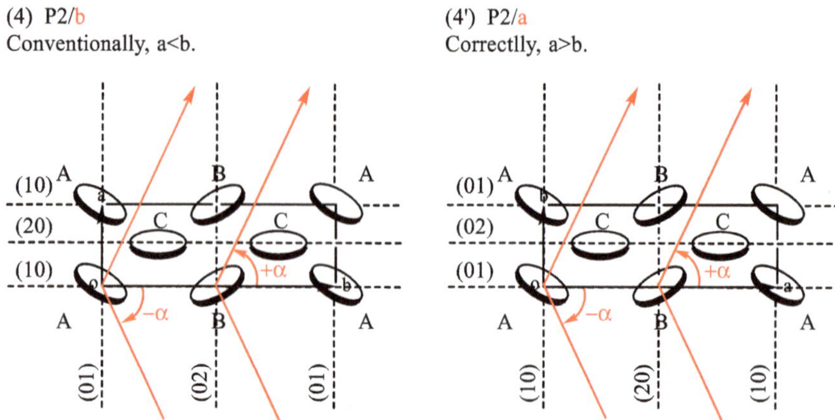

Figure 8. In these 2D rectangular lattices having P2/b and P2/a symmetries, the reflectivity is the same for both molecules A and B ($f_A = f_B$) at h = 0 for (4) and k = 0 for (4′), respectively.

When k = 0, F(hk) $= f\{1 + (-1)^k\} + f_C \cdot (-1)^h \left\{i \sin\left(\frac{\pi}{2}k\right) + i \sin\left(\frac{3\pi}{2}k\right)\right\} = 2f$: F(hk) $\neq 0$: F(hk) $\neq 0$ (appearance).

Therefore, the **extinction rule for the P2/b symmetry** can be derived as

$$\mathbf{0k: k = 2}\boldsymbol{n}\mathbf{ + 1}$$

Thus, this extinction rule is originated from the same reflectivities, $f_A = f_B$, of the molecules with the same tilt angle α, as shown in Figure 8(4): when h = 0, the A molecule on the first reflection surface (01) tilts by $-\alpha$ with respect to the b-axis, and the B molecule on the second reflection surface (02) tilted by $+\alpha$ with respect to the b-axis.

However, we have to be careful here. Conventionally, the a-axis and the b-axis have been set with being $a < b$ as shown in Figure 7(4), but this is a mistake from the nomenclature rule. Correctly, the a-axis and b-axis must be set with being $a > b$ as shown in Figure 7(4′).

In the case of a rectangular lattice with this symmetry, when k = 0, the A molecule on the first reflection surface (01) tilts by $-\alpha$ with respect to the a-axis, and the B molecule on the second reflection surface (02) tilted by $+\alpha$ with respect to the a-axis, as shown in Figure 4(4′). In this case, $f_A = f_B$. Therefore, when $a > b$,

The **extinction rule of $P2/a$ symmetry** must be

$$\mathbf{h0: h = 2}\boldsymbol{n}\mathbf{+1}$$

Therefore, since the extinction rule changes depending on whether $a < b$ or $a > b$, as above, we should take care that the a- and b-axes should be set as $a > b$, when we carry out the X-ray analysis of the liquid crystal structure with this symmetry.

Problem 6. Problems for calculation of the extinction rule when a new symmetry of two-dimensional rectangular lattice appears:

(6-1) $D_{L.rec}(P2_12_1)$ and $D_{L.rec}(P2_11)$ phases: K. Ohta, R. Higashi, M. Ikejima, I. Yamamoto and N. Kobayashi, *J. Mater. Chem.*, **8**, 1979–1991 (1998). Read this paper, and calculate the extinction rule when the symmetry of the two-dimensional lattice is $(P2_12_1)$ and $(P2_11)$.

(6-2) $M(Pa2_1)$ phase: Y. Abe, K. Nakabayashi, N. Matsukawa, H. Takashima, M. Iida, T. Tanase, M. Sugibayashi, H. Mukai, and K. Ohta, *Inorg. Chim. Acta,* **359**, 3934–3946 (2006). Read this paper and calculate the extinction rule of $(Pa2_1)$. This symmetry was a new two-dimensional rectangular lattice which had never been found till that time.

Problem (6-1).

K. Ohta, R. Higashi, M. Ikejima, I. Yamamoto and N. Kobayashi, *J. Mater. Chem.*, **8**, 1979–1991 (1998).

This paper discovered two discotic lamellar phases, $D_{L.rec}(2_11)$ and $D_{L.rec}(2_12_1)$, with novel phase structures. As shown in Figure 9, each of the metallomesonges, 4 and 5, has a bis(diphenylglyoximate)nickel complex part substituted by long alkoxy chains in the periphery. In both metallomesogens, the para-position of the phenyl group is substituted by a long-chain alkoxy group, but Complex 4 (abbreviated as Ni(n,0)) is further substituted with a hydroxy group at the meta-position, whereas Complex 5 (abbreviated as Ni(n)) is not substituted at the meta-position.

Among them, the liquid crystal phase exhibited by Complex 5: Ni(12) has a bilayer lamellar structure as shown in Figure 10. The

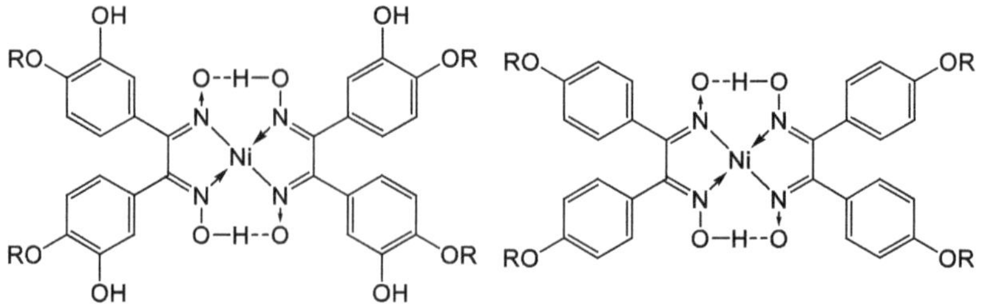

$R=C_nH_{2n+1}$, n = 12, 16
4 : Ni(n, 0)

$R=C_nH_{2n+1}$, n =10, 12, 14
5 : Ni(n)

Figure 9. Molecular structures of the bis(diphenylglyoximato)nickel(II)-based metallomesogens, 4: Ni(n, 0) and 5: Ni(n).

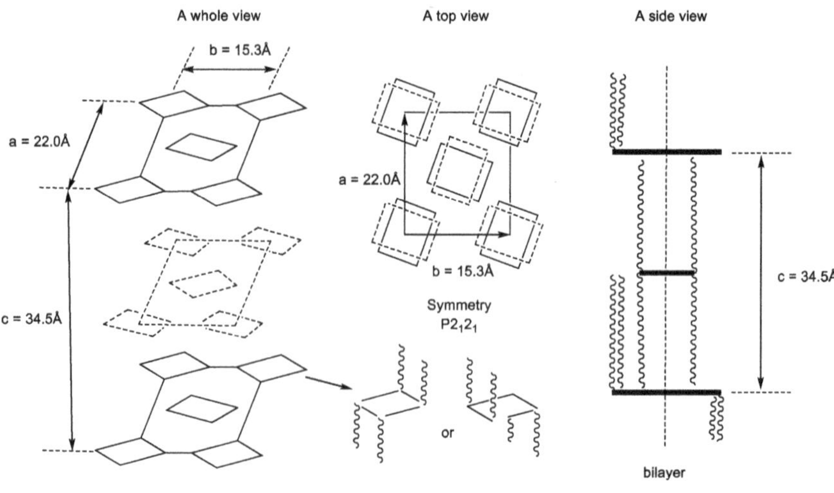

Figure 10. Structural model of the novel lamellar mesophase of $D_{L.rec.}(P2_12_1)$ of the Ni(12) complex **5**.

layer thickness of this lamellar phase is $c = 34.5\,\text{Å}$. In addition, a very characteristic feature is that the plane of the complex is perpendicular to the c-axis, which is completely different from the normal smectic liquid crystal phase, and rather similar to the discotic columnar phase. Its two-dimensional ab lattice has a $P2_12_1$ symmetry. As shown in this figure, two long-chain alkoxy groups are oriented upward and the remaining two are oriented downward,

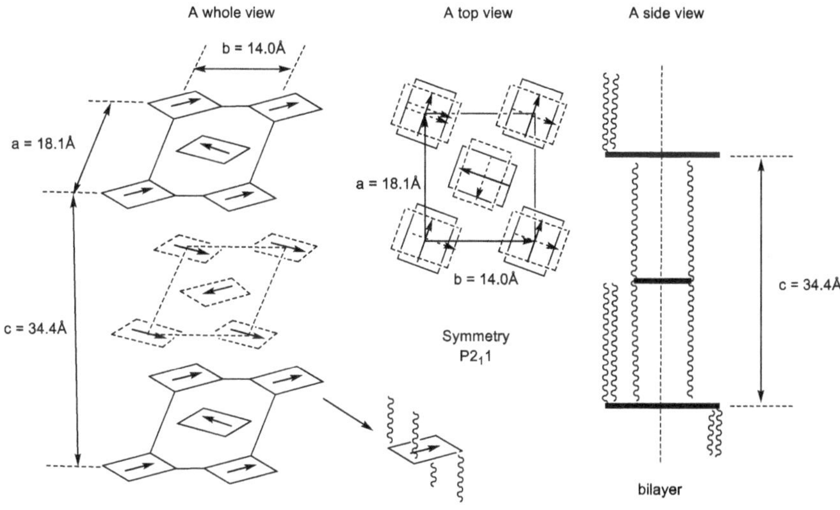

Figure 11. Structural model of the novel lamellar mesophase of $D_{L.rec.}(P2_11)$ of the Ni(12,0) complex **4**. For this structure, the (010) reflection can be observed.

giving the molecule a chair shape. The orientation of this chair is arbitrary and random for each molecule.

On the other hand, the liquid crystal phase of Complex 4: Ni(12,0) also has a bilayer lamellar structure as shown in Figure 11, and the layer thickness of this lamellar phase is $c = 34.4$ Å, which is almost the same as that of Complex 5. Furthermore, even in this liquid crystal phase, the complex plane is perpendicular to the c-axis. However, its two-dimensional ab lattice has P211 symmetry, unlike the liquid crystal phase of Complex 5. As can be seen from Figure 11, even in this liquid crystal phase, two peripheral long-chain alkoxy groups are oriented upward and the remaining two are oriented downward, giving the molecule a chair shape as a whole. The direction of the chair is fixed like an arrow. This fixation is due to the intermolecular hydrogen bonding of the hydroxyl groups at the ortho position (see the paper at the beginning of this solution).

This was found from a difference of the reflections in the X-ray diffraction between these liquid crystal phases: the (010) reflection line appears in Complex 4: Ni(12,0) but does not appear in

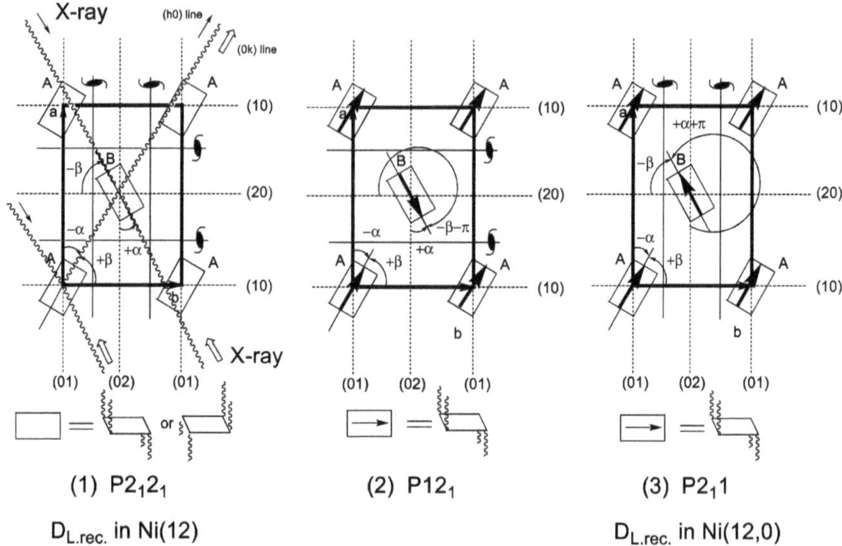

Figure 12. Structures of the rectangular lattices having $P2_12_1$, $P12_1$ and $P2_11$ symmetries.

Complex 5: Ni(12). In order to elucidate the origin of this difference, the liquid crystal phase structure factors were calculated by using three kinds of rectangular lattices as shown in Figure 12.

In the rectangular lattice shown in Figure 12(1), the orientation of the chairs is random, so there is no difference in the orientation of the chairs, and there are twofold screw axes in the a-axis direction and the b-axis direction. Therefore, the symmetry can be expressed as $P2_12_1$. On the other hand, in the rectangular lattices shown in (2) and (3) of Figure 12, the direction of the chairs is fixed, so that the symmetry is reduced. In the lattice (2), there is a twofold screw axis along the b-axis, but there is no twofold screw axis along the a-axis. Accordingly, this symmetry can be expressed as $P12_1$. In the lattice (3), there is a twofold screw axis along the a-axis, but there is no twofold screw axis along the b-axis. Accordingly, this symmetry can be expressed as $P2_11$.

Moreover, as we have already considered in Problem 5, the molecules tilted at the same angle by $+\alpha$ and $-\alpha$ with respect to a certain axis give the same reflectivity. Considering the same

Table 9. Extinction rules for the two-dimensional rectangular lattices having $P2_12_1$, $P12_1$ and $P2_11$ symmetries.

Symmetry	$P2_12_1$		$P12_1$		$P2_11$	
Type of equivalent molecule	A	B	A	B	A	B
Tilt angle of the molecule from the (h0) line	$+\beta$	$-\beta$	$+\beta$	$-\beta - \pi$	$+\beta$	$-\beta$
	$\therefore f_A = f_B = f$		$\therefore f_A \neq f_B$		$\therefore f_A = f_B = f$	
Tilt angle of the molecule from the (0k) line	$-\alpha$	$+\alpha$	$-\alpha$	$+\alpha$	$-\alpha$	$+\alpha + \pi$
	$\therefore f_A = f_B = f$		$\therefore f_A = f_B = f$		$\therefore f_A \neq f_B$	
Two-dimensional liquid crystalline structure factor: F	$F = f_A + f_B(-1)^{h+k}$		$F = f_A + f_B(-1)^{h+k}$		$F = f_A + f_B(-1)^{h+k}$	
Extinction rule [(h,k) for F = 0]	$h0 : h = 2n + 1$ $0k : k = 2n + 1$		$0k : k = 2n + 1$		$h0 : h = 2n + 1$	

reflectivity in mind, the liquid crystal structure factors can be calculated as shown in Table 9, and the corresponding extinction rules can be obtained as follows:

(1) **Rectangular lattice having $P2_12_1$ symmetry: $h0$:$h = 2n + 1$, $0k$:$k = 2n + 1$**
(2) **Rectangular lattice having $P12_1$ symmetry: $0k$:$k = 2n + 1$**
(3) **Rectangular lattice having $P2_11$ symmetry: $h0$: $h = 2n + 1$**

As already mentioned, the difference was observed in the X-ray diffraction of the liquid crystal phases that the (010) reflection line appeared in Complex 4: Ni(12,0) but it did not appear in Complex 5: Ni(12). Hereupon, when we apply the extinction rules just obtained to the present Ni complexes, we find that the (010) reflection line appears only in the rectangular lattice with $P2_11$ symmetry in

Case (3). Therefore, it can be concluded that the liquid crystalline phase of Complex 4: Ni(12,0) has $P2_11$ symmetry. On the other hand, Complex 5: Ni(12) show neither (100) nor (010), consistent to Case (1), so that we can conclude that the liquid crystalline phase has a $P2_12_1$ symmetry.

Thus, the liquid crystalline phases of Complex 4: Ni(12,0) and Complex 5: Ni(12) could be identified as $D_{L.rec.}(P2_11)$ and $D_{L.rec.}(P2_12_1)$, respectively.

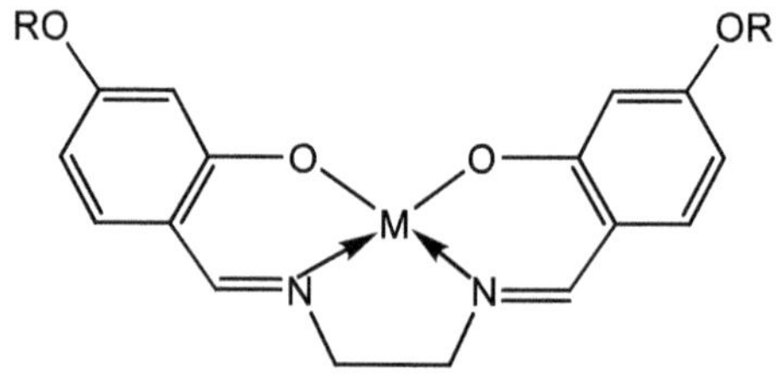

[1] M: Ni(II); R = C_nH_{2n+1} (n = 14, 16, 18, 20) Col_L

[2] M: V=O; R = C_nH2n+1 (n = 16, 18, 20) $M(Pa2_1)$ [= $Col_{L.rec.}(Pa2_1)$]

Figure 13. Molecular structure of metallomesogens based on bis(salen) metal(II) substituted by long alkoxy chains at *p*-position. The mesomorphism is strongly dependent on the central metal. Refs.: [1] Y. Abe, K. Nakabayashi, N. Matsukawa, H. Takashima, M. Iida, T. Tanase, M. Sugibayashi, H. Mukai, and K. Ohta, *Inorg. Chim. Acta,* **359**, 3934–3946 (2006). [2] Y. Abe, K. Nakabayashi, N. Matsukawa, M. Iida, T. Tanase, M. Sugibayashia, and K. Ohta, *Inorg. Chim. Commun.*, 7, 580–583 (2004).

Problem (6-2).

Very interestingly, the liquid crystal structures of the bis(salen) metal(II)-based metal complexes shown in Figure 13 change greatly depending on the type of central metal. When the central metal M is Ni(II), the liquid crystal phase is a D_L (= Col_L) phase with a dimensionality of 1D⊕1D: see Ref. [1] in the figure caption of Figure 13. On the other hand, when the central metal M is V = O, the liquid crystal phase exhibits an $M(Pa2_1)$ [= $D_{L.rec.}(Pa2_1)$] phase with [1D⊕2D] dimensionality of a rectangular lattice having an unprecedented symmetry $Pa2_1$: see Ref. [2] in the figure caption of Figure 13.

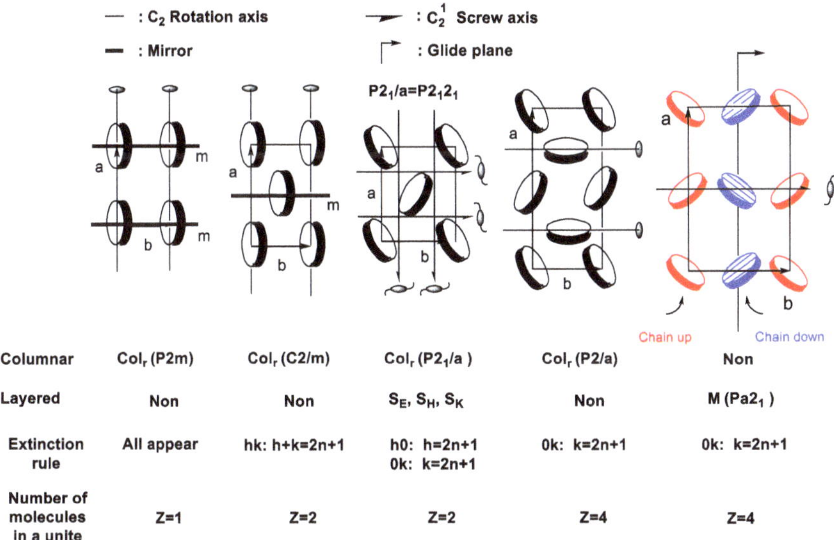

Figure 14. Relationship between packing type of cores and extinction rules for some 2D rectangular lattices appeared in mesophases.

Figure 14 summarizes the structures of columnar liquid crystal phases and layered liquid crystal phases with two-dimensional rectangular lattices known to date.

As you can see from this figure, four types of symmetries have been known so far in rectangular lattices: $P2m$, $C2/m$, $P2_1/a$, and $P2/a$. Recently, a new type of $Pa2_1$ was discovered in this paper [2] in the figure caption of Figure 13 (far right in the figure).

The X-ray crystal structure analysis of the V = O complex revealed that the two alkoxy groups extend straight in the same direction to form a layered structure, and that the central metal complex planes stand perpendicular to the layer. This crystal structure is similar to the general smectic liquid crystal structures. Normally, when a crystal phase changes to a liquid crystal phase, the alkoxy groups melt and become liquid, whereas the central cores maintain the crystal structure, so that this phase exhibits a liquid crystal phase. Therefore, the present liquid crystal phase can be considered to be classified as a smectic phase. However, in the X-ray diffraction of the liquid crystal phase, (100) and (300) reflection lines

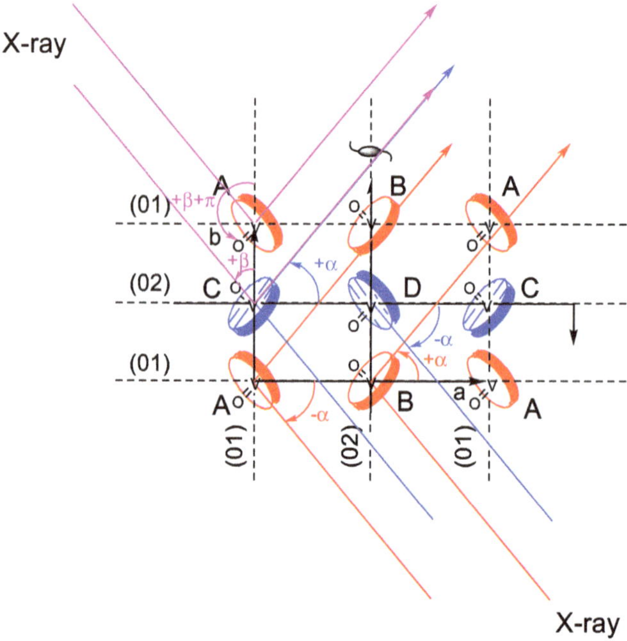

Figure 15. Schematic representation of a 2D rectangular lattice having $Pa2_1$ symmetry. Notations of α and β are the tilted angles of the molecular plane from the a- and b-axes, respectively.

appeared in this V = O complex. These reflection lines do not appear in a rectangular lattice of a conventional smectic E phase with $P2_1/a$ symmetry (see Figure 14). We wished to clarify the reason for this discrepancy, so that we calculated the liquid crystal phase structure factor and concluded that the rectangular lattice has an unprecedent symmetry of $Pa2_1$ as shown in the right end of Figure 14.

The extinction rule for this rectangular lattice was calculated as follows.

Figure 15 shows the alignment of molecules A, B, C and D which was revealed by the X-ray single crystal structure analysis. The peripheral long-chain alkoxy chains in the molecules A and B shown in red point upward from the paper surface, whereas those in molecules C and D shown in blue point downward from the paper surface. In addition, all the orientation of V = O in the core metal complex parts face leftward. Hereupon, in order to calculate

Table 10. Derivation of the extinction rule for symmetry of $Pa2_1$.

Coordinate of A	n in $nf_{\mathbf{A}}$	$\exp\{2\pi i(hx_i + ky_i)\}$
(0 0)	$\frac{1}{4}$	$\exp\{0\} = 1$
(1 0)	$\frac{1}{4}$	$\exp\{2\pi i(h)\} = 1$
(0 1)	$\frac{1}{4}$	$\exp\{2\pi i(k)\} = 1$
(1 1)	$\frac{1}{4}$	$\exp\{2\pi i(h + k)\} = 1$
Coordinate of B	n in nf_{B}	$\exp\{2\pi i(hx_i + ky_i)\}$
$\left(\frac{1}{2}0\right)$	$\frac{1}{2}$	$\exp\left\{2\pi i\left(\frac{1}{2}h\right)\right\} = (-1)^h$
$\left(\frac{1}{2}1\right)$	$\frac{1}{2}$	$\exp\left\{2\pi i\left(\frac{1}{2}h + k\right)\right\} = (-1)^h$
Coordinate of C	n in nf_{C}	$\exp\{2\pi i(hx_i + ky_i)\}$
$\left(0\frac{1}{2}\right)$	$\frac{1}{2}$	$\exp\left\{2\pi i\left(\frac{1}{2}k\right)\right\} = (-1)^k$
$\left(1\frac{1}{2}\right)$	$\frac{1}{2}$	$\exp\left\{2\pi i\left(h + \frac{1}{2}k\right)\right\} = (-1)^k$
Coordinate of D	n in nf_{D}	$\exp\{2\pi i(hx_i + ky_i)\}$
$\left(\frac{1}{2}\frac{1}{2}\right)$	1	$\exp\left\{2\pi i\left(\frac{1}{2}h + \frac{1}{2}k\right)\right\} = (-1)^{h+k}$

the liquid crystal structure factor, Table 10 was created for this rectangular lattice from Equation (10′).

Therefore, the sum of F(hk) is

$$\begin{aligned}
F(hk) &= f_A\left(\frac{1}{4}\times 1 + \frac{1}{4}\times 1 + \frac{1}{4}\times 1 + \frac{1}{4}\times 1\right) \\
&\quad + f_B\left[\frac{1}{2}\times(-1)^h + \frac{1}{2}\times(-1)^h\right] \\
&\quad + f_C\left[\frac{1}{2}\times(-1)^k + \frac{1}{2}\times(-1)^k\right] + f_D[1\times(-1)^{h+k}] \\
&= f_A + f_B\cdot(-1)^h + f_C\cdot(-1)^k + f_D\cdot(-1)^{h+k}
\end{aligned}$$

As shown in Figure 15, when we consider the case of h = 0 {*i.e.*, the reflection on (0k) surface: the primary and secondary reflections of X-rays denoted with red and blue solid lines, respectively}, A and B molecules on the (0k) reflection surface tilt by $-\alpha$ and $+\alpha$ to the a-axis, respectively. In this case, the reflectivity is $f_A = f_B$. On the other hand, in the case of C and D molecules with alkoxy groups oriented downward from the paper surface, C and D molecules on the (h0) reflection surface tilt by $-\alpha$ and $+\alpha$ to the a-axis, respectively. In this case, it also gives $f_C = f_D$. Furthermore, considering the tilt of the molecular plane from the direction of V = O, the tilt angles of all molecules of A, B, C, and D are the same $\pm\alpha$. Therefore, each of the reflectivities are the same like as $f_A = f_B = f_C = f_D$.

$$\therefore \text{F(hk)} = f_A + f_A \cdot (-1)^0 + f_A \cdot (-1)^k + f_A \cdot (-1)^{0+k} = 2[f_A + f_A \cdot (-1)^{\text{k}}]$$

From this structure factor

$$\text{When h} = 0 \text{ and k} = 2n+1,\ \text{F(hk)} = 0\,\text{(extinction)}.$$

$$\text{When h} = 0 \text{ and k} = 2n,\ \text{F(hk)} = f_A + f_B \neq 0$$

On the other hand, when we furthermore consider the case of k = 0 {*i.e.*, reflection on the (h0) surface: the primary and secondary reflections of X-rays denoted with the purple solid lines, as shown in Figure 15}, A and C molecules on the (h0) reflection surface tilt by $+\beta + \pi$ and $+\beta$ to the b-axis, respectively. In this case, the reflectivity is $f_A \neq f_C$. Similarly, in the case of B and D molecules, their reflection surfaces tilt by $+\beta$ for the B molecule and $+\beta+\pi$ for the D molecule with respect to the b-axis. In this case, the reflectivity is $f_C \neq f_D$.

$$\begin{aligned} \therefore \text{F(hk)} &= f_A + f_B \cdot (-1)^h + f_C \cdot (-1)^0 + f_D \cdot (-1)^{h+0} \\ &= f_A + f_B \cdot (-1)^h + f_C + f_D \cdot (-1)^h \end{aligned}$$

From this structure factor

$$\text{When } k = 0, h = 2n+1, \text{F(hk)} = f_A - f_B + f_C - f_D \neq 0$$

$$\text{When } k = 0, h = 2n, \quad \text{F(hk)} = f_A + f_B + f_C + f_D \neq 0$$

Thus, **the extinction rule for the $Pa2_1$ symmetry** is

$$\mathbf{0k: k = 2}\boldsymbol{n}\mathbf{+1}$$

From this extinction rule, we can theoretically understand that the liquid crystal phase with this symmetry does not give the X-ray reflection lines of (100) and (300).

Problem 7. When a disk-like molecule $C_{12}PzH_2$ was heated to 150°C, the X-ray diffraction lines were obtained as follows:

Spacing d (Å)	Intensity	line width
27.6	Large	Sharp
15.9	Medium	Sharp
13.3	Small	Sharp
ca. 4.7	Medium	Broad
ca. 3.5	Medium	Broad

Identify this liquid crystal phase by using "Reciprocal Lattice Method." Also calculate the lattice constants.

When you want to estimate the structure of the liquid crystal phase from the X-ray diffraction pattern, we can utilize the following "**Four golden rules for liquid crystal structure analysis**" proposed by Ohta in the previous textbook. (See details in Problem 9.)

Following this procedure, you will be able to identify the liquid crystal phase of $C_{12}PzH_2$ in this Problem.

1. We carry out at first the following calculations in order to estimate the liquid crystalline phase structure by employing the golden rules sequentially from Article 1.

$$27.6\,\text{Å} \div 2 = 13.6\,\text{Å}$$

$$27.6\,\text{Å} \div \sqrt{3} = 15.9\,\text{Å}$$

$$27.6\,\text{Å} \div \sqrt{2} = 19.5\,\text{Å}$$

When we compare these calculated values with the observed values, we notice that the calculated values of 13.6 Å and 15.9 Å correspond to the observed values of 13.3 Å and 15.9 Å. Therefore, we can presume this liquid crystalline phase as a Col_h phase because Article 2 can be adapted.

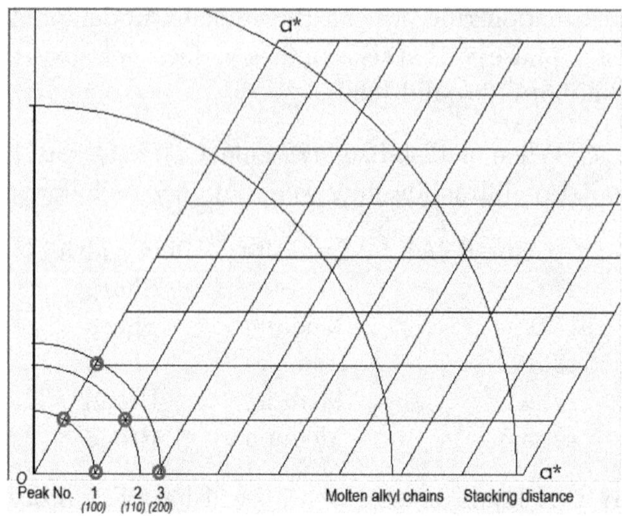

Figure 16. The reciprocal lattice of two-dimensional hexagonal lattice for $C_{12}PzH_2$ at 150°C.

2. We assume that the reflection on the lowest angle side is (hk) = (10). Therefore, $d_{10} = 27.6\,\text{Å}$.

   Since this phase has a two-dimensional hexagonal, the lattice constant a can be calculated from formula 4-1:

$$\frac{1}{d_{hk0}^2} = \frac{4}{3}\left(\frac{h^2+hk+k^2}{a^2}\right) \tag{4-1}$$

$$\frac{1}{27.6^2} = \frac{4}{3} \times \frac{1}{a^2}$$

$$\therefore a = 27.6 \times 2 \div \sqrt{3} = 31.9\,\text{Å}$$

The relationship between the real lattice and the reciprocal lattice of the 2D hexagonal lattice is $a^* = \frac{2}{\sqrt{3}a}$, $\gamma^* = 60°$. Therefore,

$$a^* = \frac{2}{\sqrt{3} \times 31.9} = 0.0362 \xrightarrow{\times 50} 1.81\,cm$$

A reciprocal lattice plane with $a^* = 1.81\,cm$, $\gamma^* = 60°$ was drawn in Figure 16. In addition, we calculated the reciprocal radii from

Table 11. Calculation of the radii of the observed d values.

Peak No.	1	2	3	4	5
d(Å)	27.6	15.9	13.3	4.7	3.5
$(1/d) \times 10^{-2}$	3.62	6.29	7.51	21.3	28.6
$\times 50$ cm	1.81	3.15	3.76	10.6	14.3

the observed spacing d values (Table 11) and the quarter circles were drawn in this reciprocal lattice plane.

On this reciprocal lattice plane, it can be seen that the Debye-Scherrer rings of Peak Nos. 1 to 3 agree well with the lattice intersection points. When you read out the intersection (hk) points, you can index these three reflections as (10), (11) and (20). Furthermore, Peak No. 4 is a very broad and large peak, corresponding to the average distance between the fused alkyl groups. Peak No. 5 located at d = 3.5 Å corresponds to the (001) reflection of the stacking distance h between discotic molecules in the column.

From the above-mentioned reciprocal lattice analysis method, the liquid crystalline phase of $C_{12}PzH_2$ can be identified as a hexagonal ordered columnar (Col_{ho}) phase with lattice constants a = 31.9 Å and h = 3.5 Å.

3. There is no extinction rule for a 2D hexagonal lattice, so that all reflection lines appear. Therefore, no extinction rule is applied here.
4. Next, we calculate the Z value. The density $\rho(g/cm^3)$ of the liquid crystal phase is represented by

$$\rho = (ZM)/(VN) \tag{17}$$

Z (number): number of molecules in a unit cell,

M (g/mol) : molecular weight,

V (cm^3) : volume of unit cell,

N (number/mol): Avogadro's number

Therefore,

$$Z = (\rho V N)/M \tag{18}$$

Although it is extremely difficult to measure the density of the liquid crystal phase, it is usually $0.8 \sim 1.2(g/cm^3)$. In addition, the unit cell volume V in each of the liquid crystalline phases of Col_h, Col_r, Col_{tet}, and Col_{ob}, can be colligatively presented as follows:

$$V = abc \sin\gamma \tag{19}$$

When a Col_h phase, $a = b$.

Hereupon, we consider the stacking distance of the columnar liquid crystalline phase. The one-dimensional stacking distance in this column is along the c-axis, so that $d_{001} = c$ (lattice constant). If the columnar liquid crystalline phase is an ordered phase with the dimension of [2D $\oplus$ 1D], $d_{001} = c$ can be obtained immediately. However, d_{001} does not appear in the disordered columnar phases like as Col_{hd}, $Col_{tet.d}$, $Col_{ob.d}$, *etc.*, which have only 2D dimensions. However, we can assume $c = 3.3$ to 3.5 Å from the van der Waals distance of general organic discotic molecules. In addition, the discotic molecules in the case of Col_r phase are tilted with respect to the column axis, the stacking distance (c) in this column can be assumed 4 to 6 Å. These assumptions are only a guideline. We are required to pay very careful attention: there is a Col_r phase with tilted disks even for $c = 3.4$ Å, and there is a dimer-forming Col_h phase with non-tilted disks even for $c = 7$ Å.

Here, we can consider $c = ca.3.5$ Å from the sentence in this problem.

Therefore, the Z value can be calculated as follows:

$$Z = (\rho V N)/M = \frac{\rho(abc\,\sin\gamma)N}{M} \qquad (\because (18),(19))$$

$$= \frac{(1.0\,g/cm^3) \times (31.9 \times 10^{-8}\,cm)^2 \times (3.5 \times 10^{-8}\,cm) \times \sin 120^\circ \times (6.02 \times 10^{23}\,number\,of\,molecules/mol)}{1869.05\,g/mol}$$

$$= 0.99 \cong 1(number\,of\,molecules)$$

The Z value obtained from formula (18) should always be an integer. If it does not become an integer, the identification is erroneous. The presently calculated value 0.99 is not perfect integer. This is because the density is assumed to be $1.0\,g/cm^3$, and because the peak at $c = ca.\,3.5\mathring{A}$ is broad with considerable fluctuations. Since $Z \cong 1$, it supports that the identification of the liquid crystalline phase of $C_{12}PzH_2$ as the Col_h phase is correct.

Thus, this liquid crystalline phase could be identified as a Col_h phase having the lattice constants: $a = b = 31.9$ Å and $c = ca.\,3.5$ Å.

Problem 8. (8-1) $C_{12}PzCu$ could be prepared by metallation of the metal-free derivative $C_{12}PzH_2$ in Problem 7. When this Cu(II) complex was heated to 150°C, the X-ray diffraction lines were obtained as follows:

Spacing d (Å)	intensity	line width
28.9	Large	Sharp
26.3	Large	Sharp
16.4	Medium	Sharp
14.7	Small	Sharp
13.2	Small	Sharp
ca. 4.7	Medium	Broad
ca. 3.4	Medium	Broad

Identify this liquid crystal phase by using "Reciprocal Lattice Method." Also calculate the lattice constants.

(8-2) Considering the extinction rules for 2D lattices, determine the lattice symmetry of this mesophase.

We resolve this problem using the same procedure as for Problem 7.

(8-a)

1. At first, we carry out the following calculations in order to estimate the liquid crystalline phase structure by employing the

golden rules sequentially from Article 1.

$$28.9\,\text{Å}/2 = 14.5\,\text{Å}$$

$$28.9\,\text{Å} \div \sqrt{3} = 16.7\,\text{Å}$$

$$28.9\,\text{Å}/\sqrt{2} = 20.4\,\text{Å}$$

When we compare these calculated values with the observed values.

(1) Estimation from Article 1: the calculated value of 14.5 Å agrees with the observed value of 14.7 Å within the range of measurement error, but the values of 26.3 Å and 16.4 Å cannot be explained only from the lamellar ratio.
(2) Estimation from Article 2: the calculated value of 16.7 Å agrees with the observed value of 16.4 Å within the range of measurement error, but the observed value of 26.3 Å cannot be explained. Therefore, it is not a 2D hexagonal lattice. Moreover,
(3) Estimation from Article 3: the calculated value of 20.4 Å does not correspond to the observed value at all. Therefore, a 2D-tetragonal lattice cannot be considered. Therefore, it is neither 1D-lamellar nor 2D-hexagonal nor 2D-tetragonal, so it is thought to have a 2D-rectangular lattice or a 2D-oblique lattice.
(4) When we calculate the ratios from the observed values, we obtain the following ratios:

$$28.9\text{Å} : 14.7\text{Å} = 1 : 1/2$$

$$26.3\text{Å} : 13.2\text{Å} = 1 : 1/2$$

Thus, it seems that there are two series of lamellar type ratios. This corresponds to Article 4 and is characteristic to a 2D-rectangular lattice.
From the above, it can be inferred that it is a Col_r phase.

2. Since the strongest intensity of reflections from a 2D-rectangular lattice can be expected from the (20) and (11) planes, we calculate

the lattice constants from Equation (2-1) assuming as d_{20} = 28.9 Å, d_{11} = 26.3 Å.

$$\frac{1}{28.9^2} = \frac{2^2}{a^2} + \frac{0^2}{b^2}$$
$$\frac{1}{26.3^2} = \frac{1^2}{a^2} + \frac{1^2}{b^2}$$

When we solve these two equations simultaneously, we get the a and b values as follows:

$$a = 57.8\,\text{Å},$$
$$b = 29.5\,\text{Å}$$

Therefore, the length of two sides of the reciprocal lattice is calculated as follows:

$$a^* = \frac{1}{a} = \frac{1}{57.8} = 0.0173 \xrightarrow{\times 50} 0.865\,\text{cm}$$
$$b^* = \frac{1}{b} = \frac{1}{29.5} = 0.0339 \xrightarrow{\times 50} 1.70\,\text{cm}$$

A reciprocal lattice plane with a^* = 0.865 cm, b^* = 1.70 cm, $\gamma = 90°$ was drawn in (Figure 17). In addition, the quarter circles. In addition, we calculated the reciprocal radii from the observed spacing d values (Table 12) and the quarter circles were drawn in this reciprocal lattice plane.

Table 12. Calculation of the radii of the observed d values.

Peak No	1	2	3	4	5	6	7
d(Å)	28.9	26.3	16.4	14.7	13.2	4.7	3.4
(1/d) × 10^{-2}	3.46	3.80	6.10	6.89	7.58	21.3	29.4
×50 cm	1.73	1.90	3.05	3.40	3.78	10.6	14.7

Peak Nos. 6 and 7 in Table 12 are broad peaks, which can be assigned as the average distance of melted peripheral long alky chains and the average inter-disc distance of discotic molecules stacked in

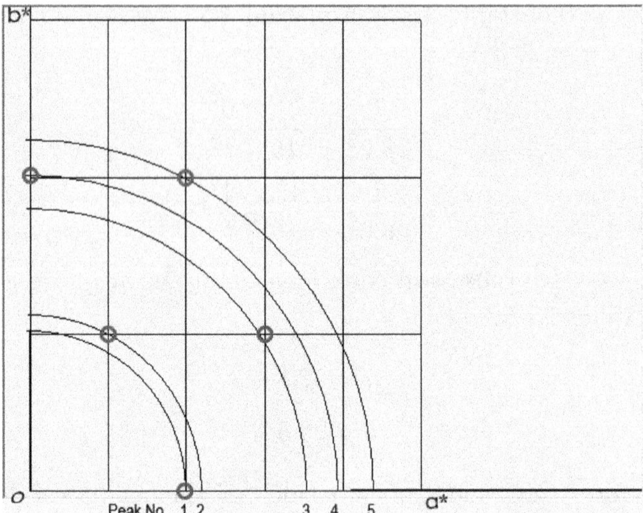

Figure 17. The reciprocal lattice of rectangular lattice for C_{12}PzCu at 150°C.

columns (stacking distance $= c$), respectively. Therefore, the peaks related to two-dimensional column packing are Nos. 1–5.

We index the reflection peaks from the intersection positions of the reciprocal lattice and the quarter circles in Figure 17. Since (20) and (11) are taken as the standard peaks, it is natural that they pass through the lattice points. The other three circles are well matched to intersection points (31), (02), and (22), so that we can conclude that our first assumption was correct. Therefore, this liquid crystal phase can be identified as a $\mathrm{Col_r}$ phase.

(8-b)

3. Since there is no reflected line of $\mathrm{h} + \mathrm{k} = 2n + 1$, the symmetry of this 2D-rectangular lattice can be identified as C2/m from Equation (11).
4. In order to confirm that the above identification is correct, Z value calculation is carried out. In this 2D-rectangular lattice, $\gamma = 90°$. Therefore, the Z value can be calculated from Equations (18) and

(19) as follows:

$$Z = \frac{(1.0\,g/cm^3) \times (57.8 \times 10^{-8}\,cm) \times (29.5 \times 10^{-8}\,cm) \times (3.4 \times 10^{-8}\,cm) \times (6.02 \times 10^{23}\,number/mol)}{1930.58\,g/mol}$$

$$= 1.8 \cong 2(\text{number of molecules})$$

Since the calculated value $Z \cong 2$ is consistent with the theoretical value $Z = 2$ for $Col_r(C2/m)$, we can confirm that the identification of this liquid crystal phase as $\mathrm{Col_r}(C2/m)$ is correct.

From the above, this liquid crystal phase is assigned as a Colr$(C2/m)$ phase with lattice constants: a = 57.8 Å, b = 29.5 Å, and c = *ca.*3.4 Å.

Problem 9. The attached sheets are a list of X-ray diffraction data for discotic liquid crystal phases. Choose ten favorite data from this list and identify each liquid crystal phase. Be sure to identify all the columnar phases by using the reciprocal lattice method other than a simple discotic lamellar phase. Give the Miller index for each diffraction line and calculate the lattice constants. If necessary, determine the symmetry of the lattice.

The structural analysis of the liquid crystal phase needs to be performed in the following order.

① Estimation of n-dimensional lattice by "golden rule for liquid crystal structure analysis"
② Analysis by reciprocal lattice and indexing
③ Application of the extinction rule
④ Verification by Z value calculation

Next, when we actually analyze the structure of liquid crystalline phase, we sequentially employ the following guideline of four "**Golden rules for liquid crystal structure analysis**" proposed by Ohta.

Golden rule of liquid crystal structure analysis

① If the liquid crystal phase has a one-dimensional layered (lamella) structure, from the Equation (3-3) in the textbook, the ratio of spacing d will be given as follows:

$$d_1 : d_2 : d_3 : d_4 : d_5 : \cdots = 1 : \frac{1}{2} : \frac{1}{3} : \frac{1}{4} : \frac{1}{5} : \cdots \tag{3}$$

② If the liquid crystal phase has a two-dimensional hexagonal structure, from the Equation (3-5) in the textbook, the ratio of spacing d will be given as follows:

$$d_1 : d_2 : d_3 : d_4 : d_5 : d_6 : \cdots = 1 : \frac{1}{\sqrt{3}} : \frac{1}{2} : \frac{1}{\sqrt{7}} : \frac{1}{3} : \frac{1}{\sqrt{12}} : \cdots \tag{5}$$

③ If the liquid crystal phase has a two-dimensional tetragonal structure, from the Equation (3-8) in the textbook, the ratio of spacing d will be given as follows:

$$d_1 : d_2 : d_3 : d_4 : d_5 : d_6 : \cdots = 1 : \frac{1}{\sqrt{2}} : \frac{1}{2} : \frac{1}{\sqrt{5}} : \frac{1}{3} : \frac{1}{\sqrt{8}} : \cdots \tag{8}$$

④ If the liquid crystal phase has a two-dimensional rectangular structure, it cannot be represented by a simple ratio such as the above-mentioned articles from 1 to 3. However, we sometimes notice that the ratio shows two series of layer structures (the nature of article 1). This is the feature of the rectangular structure.

When this phase is likely to be a rectangular phase, the two strong reflections in the lowest angle region are assumed as the reflections of (20) and (11) (or (11) and (20)). By substituting the spacing values of d_{20} and d_{11}, the lattice constants, a and b, are calculated from the following equation:

$$\frac{1}{d_{hk0}^2} = \frac{h^2}{a^2} + \frac{k^2}{b^2} \tag{2-1}$$

You draw a reciprocal lattice by using these obtained lattice constants, a and b, and carry out the analysis. Then, you use the extinction rules (Fig. 12 in the textbook) to determine the symmetry of the lattice.

If you know the above articles from (1) to (4), you can analyze most of the structures in a wide range of soft matters such as rod-like liquid crystals, discotic liquid crystals, thermotropic liquid crystals, lyotropic liquid crystals, low-molecular-weight liquid crystals, and high-molecular-weight liquid crystals.

Analysis of liquid crystal phase by "Reciprocal lattice method"

At first, we calculate the ratios from the spacing d values obtained from the X-ray diffraction of the liquid crystal phase. Next, we sequentially employ the above-mentioned four "golden rules of liquid crystal structure analysis" in order to estimate what dimension (1D or 2D) of a lattice the liquid crystal phase has.

If the present liquid crystal phase exhibits only a one-dimensional lattice corresponding to Article 1, it can be analyzed by a very simple ratio. From this ratio alone, it can be identified as a lamellar type liquid crystal (S_A, S_C, *etc.*). By the way, S_A and S_C cannot be distinguished from each other basically by X-ray structural analysis, so that we distinguish them from the difference in texture by using a polarizing microscope.

On the other hand, when the present liquid crystal has a two-dimensional lattice (corresponding to Articles 2 to 4), the method of identifying only from the ratios is not recommended due to many misidentifications. This analysis methodology has been performed all over the world up to date, but there are many misidentifications to the wrong phases, and even when the phase is identified as the correct phase, the incorrect indexation is undertaken in many cases.

Therefore, in the case of a two-dimensional lattice, Reciprocal lattice method is strongly recommended instead of Ratio method.

The relationship between the two-dimensional real and reciprocal lattices are shown in Table 11 of the textbook, so that we refer to it.

There are only the following four two-dimensional lattices of liquid crystal (see Table 6 in the textbook):

2D hexagonal lattice ($\mathrm{Col_h}$, $\mathrm{S_B}$, *etc.*),
2D tetragonal (square) lattice ($\mathrm{Col_{tet}}$, $\mathrm{S_T}$),
2D rectangular lattice ($\mathrm{Col_r}$, $\mathrm{S_E}$, *etc.*),
2D oblique lattice ($\mathrm{Col_{ob}}$, $\mathrm{S_F}$, S_I *etc.*).

Therefore, if we know only these four lattices, we can analyze the structure of almost all liquid crystal phases. However, since we rarely encounter the two-dimensional oblique lattice, it is enough to know first three of them.

According to the above strategy, we analyze the liquid crystal phase structure from each set of the observed X-ray diffraction spacing values listed in a table of Problem 9, by using **Reciprocal lattice method**.

Basically, if you have a calculator, graph paper, a ruler, and a compass, you can analyze the liquid crystal phase structure by the reciprocal lattice method. The analysis results are summarized in Table 13 below.

Table 13. X-ray data and the identifications of mesophases for Sample (1)~(114). The liquid crystalline phases were analyzed by "Reciprocal Lattice Method" and /or "FlexiLattice Method" developed by us.

z	○ (1)Yamaguchi1			(2)Yamaguchi2			(3)Yamaguchi3		
Reference	p68: $(C_{12}O)_{16}$-TTPH$_2$(1)			p68: $(C_{12}O)_{16}$-TTPH$_2$(1)			p68: $(C_{12}O)_{16}$-TTPCu (1-Cu)		
Mw	Mw=4172.74			Mw=4172.74			Mw=4234.21		
Temperature	RT			55°C			RT		
Mesophase	Col$_{ho}$			Col$_{ho}$			Col$_{ho}$		
Lattice constants	a = 41.5, Z = 2.0 (dimer) for ρ = 0.95			a = 39.7, h = *ca.* 9.1, Z = 2.0 (dimer) for ρ = 1.1			a = 41.9, h = ca. 9.3, Z = 2.0 (dimer) for r = 1.0		
Peak No.	$d_{obs.}$	$d_{calcd.}$	*(hkl)*	$d_{obs.}$	$d_{calcd.}$	*(hkl)*	$d_{obs.}$	$d_{calcd.}$	*(hkl)*
1	36.0	36.0	(1 0 0)	34.4	34.4	(1 0 0)	36.3	36.3	(1 0 0)
2	21.4	20.8	(1 1 0)	19.9	19.9	(1 1 0)	21.2	20.9	(1 1 0)
3	9.55	sharp	(0 0 1)	ca.9.1	broad	(0 0 1)	ca.9.3	broad	(0 0 1)
4	ca.4.3	–	#	ca.4.4	–	#	ca.4.4	–	#

z	(4)Yamaguchi4			(5)Yamaguchi5			(6)Yamaguchi6		
Reference	p68: $(C_{12}O)_8$-BTPH$_2$ (4)			p68: $(C_{16}O)_8$-BTPH$_2$ (4)			p68: $(C_{16}O)_8$-BTPH$_2$ (4)		
Mw	Mw=2241.54			Mw=2690.38			Mw=2690.38		
Temperature	125°C			60°C			120°C		
Mesophase	Col$_{rd}$(C2/m)			Col$_{rd}$(P2$_1$/a)			Col$_{rd}$(C2/m)		
Lattice constants	a = 59.7, b = 40.0			a =74.1, b = 39.4			a =72.6, b = 51.0		
Peak No.	$d_{obs.}$	$d_{calcd.}$	*(hkl)*	$d_{obs.}$	$d_{calcd.}$	*(hkl)*	$d_{obs.}$	$d_{calcd.}$	*(hkl)*
1	33.2	33.2	(1 1 0)	37.0	37.0	(2 0 0)	41.7	41.7	(1 1 0)
2	29.8	29.8	(2 0 0)	34.8	348	(1 1 0)	36.3	36.3	(2 0 0)
3	16.6	16.6	(2 2 0)	18.8	19.1	(1 2 0)	22.0	21.9	(3 1 0)
4	ca.4.4	–	#	13.4	13.5	(4 2 0)	10.5	10.4	(4 4 0)
5				10.7	10.7	(4 3 0)	ca.4.6	–	#
6				4.41	–	–[b]			
7				4.14	–	–[b]			
8									
9									
10									
11									
12									
13									
14									
15									

#: halo of molten alkyl chains

b: see the main text of p68.

Definition of the sufix discription for the mesophases:

o: When the stacking distance could be observed in any magnitude, the mesophase was assigned an **ordered** phase.

d: Only when the stacking distance could not be observed at all, the mesophase was assigned a **disordered** phase.

Table 13. (*Continued*)

Entry No.	(7)Ikejima1			◎ (8)Oka1			(9)Azumane1			(10)Azumane2			(11)Azumane3			(12)Azumane4		
Reference	p52, p66, (but no XRD data therein),$[(C_{12}O)_4DPG]_2Pt$			p75: $[(C_5O)_4DPD]_2Pd$			p60, 4a: $[(C_{10}O)_2Ph]_8PcCu$;2 Master Thesis			p60, 4a: $[(C_{10}O)_2Ph]_8PcCu$;2 Master Thesis			p60, 4b, C_{11}-Cu, $[(C_{11}O)_2Ph]_8PcCu$;3 Master Thesis			p60, 4b, C_{11}-Cu, $[(C_{11}O)_2Ph]_8PcCu$;3 Master Thesis		
Mw	Mw=2148.16			Mw=1280.22			Mw=3685.15			Mw=3685.15			Mw=3909.62			Mw=3909.62		
Temperature	150°C			130°C			100°C			212°C			100°C			201°C		
Mesophase	Col_{hd}			Col_{ho}			$Col_{rd}(P2_1/a)$			$Col_{rd}(P2_1/a)$			$Col_{rd}(P2_1/a)$			$Col_{rd}(P2_1/a)$		
Lattice constants	a = 34.4			a = 25.1, h = 3.38; Z = 1.0 for ρ = 1.2			a = 56.6, b = 38.8			a = 54.0, b = 39.5			a = 63.1, b = 43.8			a = 59.1, b = 43.9		
Peak No.	$d_{obs.}$	$d_{calcd.}$	(*hkl*)	$d_{obs.}$	$d_{calcd.}$	(*hkl*)	$d_{obs.}$	$d_{calcd.}$	(*hkl*)	$d_{obs.}$	$d_{calcd.}$	(*hkl*)	$d_{obs.}$	$d_{calcd.}$	(*hkl*)	$d_{obs.}$	$d_{calcd.}$	(*hkl*)
1	28.0	29.8	(1 0 0)	21.6	21.7	(1 0 0)	32.0	32.0	(1 1 0)	31.1	31.9	(1 1 0)	36.0	36.0	(1 1 0)	33.4	35.2	(1 1 0)
2	16.7	17.2	(1 1 0)	12.6	12.5	(1 1 0)	28.3	28.3	(2 0 0)	27.2	27.0	(2 0 0)	31.5	31.5	(2 0 0)	29.1	29.1	(2 0 0)
3	14.7	14.9	(2 0 0)	10.9	10.9	(2 0 0)	18.6	18.4	(1 2 0)	19.8	19.8	(0 2 0)	19.7	19.0	(3 1 0)	24.5	24.5	(2 1 0)
4	11.2	11.3	(2 1 0)	8.20	8.21	(2 1 0)	12.6	12.6	(1 3 0)	11.8	11.8	(2 3 0)	17.8	18.0	(2 2 0)	22.0	22.0	(0 2 0)
5	9.87	9.94	(3 0 0)	7.25	7.24	(3 0 0)	11.2	11.2	(4 2 0)	ca.4.8	–	#	13.1	13.2	(2 3 0)	18.5	18.0	(3 1 0)
6	8.54	8.61	(2 2 0)	6.28	6.27	(2 2 0)	ca.4.8	–	#				9.65	9.71	(3 4 0)	15.0	14.8	(4 0 0)
7	8.22	8.28	(3 1 0)	6.01	6.02	(3 1 0)							8.58	8.53	(6 3 0)	12.8	13.1	(2 3 0)
8	7.46	7.46	(4 0 0)	5.41	5.43	(4 0 0)							ca.4.8	–	#	ca.4.8	–	#
9	6.83	6.84	(3 2 0)	ca.4.5	–	#												
10	6.50	6.51	(4 1 0)	3.38	–	h: (0 0 1)												
11	5.94	5.97	(5 0 0)															
12	5.64	5.64	(4 2 0)															
13	5.34	5.36	(5 1 0)															
14	ca.4.5	–	#															
15																		

(*Continued*)

Table 13. (*Continued*)

Entry No.	(13)Azumane5			(14)Azumane6			○ (15)Azumane7			(16)Hatsusaka11			(17)Azumane8			○ (18)Azumane9		
Reference	p60, 4c ,C_{12}-Cu $[(C_{12}O)_2Ph]_8PcCu$;4 Master Thesis			p60, 4c ,C_{12}-Cu $[(C_{12}O)_2Ph]_8PcCu$;4 Master Thesis			p74, 1c, $(C_{12}OPh)_8PzCu$ = $(C_{12}O)_8$-Cu			p84, $[(C_{14}O)_2PhO]_8PcCu$ (8f)			p74, 1c, $(C_{12}OPh)_8PzCu$ = $(C_{12}O)_8$-Cu			p74, 2c, $(C_{12}OPh)_8PzNi$ = $(C_{12}O)_8$-Ni		
Mw	Mw=4134.01			Mw=4134.01			Mw=2667.36			Mw=4710.87			Mw=2667.36			Mw=2662.52		
Temperature	100°C			196°C			90°C			147°C			180°C			90°C		
Mesophase	$Col_{rd}(P2_1/a)$			$Col_{rd}(P2_1/a)$			$Col_{tet.o}$			$Col_{tet.d}$			$Col_{rd}(P2_1/a)$			$Col_{tet.o}$		
Lattice constants	a = 61.0, b = 42.2			a = 60.2, b = 42.6			a = 31.3, h = 3.77, Z = 1.0 for ρ = 1.2			a = 45.1			a = 68.0, b = 29.9			a = 30.7, h = ca. 3.7, Z = 0.99 for ρ = 1.25		
Peak No.	$d_{obs.}$	$d_{calcd.}$	*(hkl)*	$d_{obs.}$	$d_{calcd.}$	*(hkl)*	$d_{obs.}$	$d_{calcd.}$	*(hkl)*	$d_{obs.}$	$d_{calcd.}$	*(hkl)*	$d_{obs.}$	$d_{calcd.}$	*(hkl)*	$d_{obs.}$	$d_{calcd.}$	*(hkl)*
1	34.7	34.7	(1 1 0)	34.8	34.8	(1 1 0)	31.3	31.3	(1 0 0)	31.9	31.9	(1 1 0)	34.0	34.0	(2 0 0)	32.0	30.7	(1 0 0)
2	30.5	30.5	(2 0 0)	30.1	30.1	(2 0 0)	23.0	22.1	(1 1 0)	22.7	22.7	(2 2 0)	27.4	27.4	(1 1 0)	22.6	21.7	(1 1 0)
3	17.7	17.4	(2 2 0)	24.7	24.6	(2 1 0)	11.2	11.1	(2 2 0)	9.09	9.02	(3 4 0)	17.1	17.0	(4 0 0)	10.2	10.2	(3 0 0)
4	15.8	15.3	(4 0 0)	21.9	21.3	(0 2 0)	10.4	10.4	(3 0 0)	ca.4.6	–	#	13.8	13.7	(2 2 0)	7.59	7.68	(4 0 0)
5	13.6	13.7	(1 3 0)	13.6	13.2	(3 1 0)	7.41	7.38	(3 3 0)				12.1	12.4	(5 1 0)	7.14	7.24	(3 3 0)
6	12.1	12.4	(4 2 0)	14.9	15.1	(4 0 0)	ca.4.8	–	#				ca.4.8	–	#	ca.4.8	–	#
7	9.94	9.97	(2 4 0)	13.0	12.9	(2 3 0)	3.77	–	(0 0 1)							ca.3.7	–	h: (0 0 1)
8	8.83	8.68	(4 4 0)	11.5	11.6	(3 3 0)												
9	7.95	7.98	(5 4 0)	ca.4.8	–	#												
10	ca.4.8	–	#															
11																		
12																		
13																		
14																		
15																		

(*Continued*)

Table 13. (*Continued*)

Entry No.	○ (19)Azumane10			◎ (20)Azumane11			○ (21)Azumane12			○ (22)Azumane13			◎ (23)Azumane14			○ (24)Azumane15		
Reference	p74, 2c, $(C_{12}OPh)_8PzNi$			p74, 1b, $(C_{10}OPh)_8PzCu = (C_{10}O)_8\text{-}Cu$			p74, 3c, $[(C_{12}O)_2Ph]_8PzCu = (C_{12}O)_{16}\text{-}Cu$			p74, 3c, $[(C_{12}O)_2Ph]_8PzCu = (C_{12}O)_{16}\text{-}Cu$			p74, 3c, $[(C_{12}O)_2Ph]_8PzCu = (C_{12}O)_{16}\text{-}Cu$			p74, 4c, $[(C_{12}O)_2Ph]_8PzNi = (C_{12}O)_{16}\text{-}Ni$		
Mw	Mw=2662.52			Mw=2442.93			Mw=4141.95			Mw=4141.95			Mw=4141.95			Mw=4137.11		
Temperature	150°C			140°C			RT			150°C			210°C			RT		
Mesophase	$Col_{rd}(C2/m)$			$Col_{tet.o}$			Col_{hd}			$Col_{rd}(C2/m)$			$Col_{rd}(C2/m)$			Col_{hd}		
Lattice constants	a = 53.3, b = 41.9			a = 29.95, h = 3.82; Z = 1.0 for ρ = 1.20			a = 34.8			a = 71.6, b = 33.2			a =72.4, b = 33.0			a = 36.3		
Peak No.	$d_{obs.}$	$d_{calcd.}$	*(hkl)*	$d_{obs.}$	$d_{calcd.}$	*(hkl)*	$d_{obs.}$	$d_{calcd.}$	*(hkl)*	$d_{obs.}$	$d_{calcd.}$	*(hkl)*	$d_{obs.}$	$d_{calcd.}$	*(hkl)*	$d_{obs.}$	$d_{calcd.}$	*(hkl)*
1	32.9	32.9	(1 1 0)	29.7	29.9	(1 0 0)	30.1	30.1	(1 0 0)	35.8	35.8	(2 0 0)	36.2	36.2	(2 0 0)	31.5	31.5	(1 0 0)
2	26.6	26.6	(2 0 0)	21.4	21.2	(1 1 0)	16.9	17.4	(1 1 0)	30.1	30.1	(1 1 0)	30.0	30.0	(1 1 0)	18.4	18.2	(1 1 0)
3	13.5	13.5	(1 3 0)	9.39	9.47	(1 3 0)	15.1	15.1	(2 0 0)	19.6	19.4	(3 1 0)	19.3	19.5	(3 1 0)	8.72	8.72	(3 1 0)
4	10.9	10.9	(3 3 0)	8.30	8.31	(2 3 0)	11.2	11.4	(2 1 0)	16.7	16.6	(0 2 0)	14.9	15.0	(2 2 0)	ca.4.8	–	#
5	ca.4.8	–	#	7.27	7.26	(4 1 0)	9.93	10.0	(3 0 0)	15.0	15.1	(2 2 0)	12.3	12.2	(4 2 0)			
6				6.78	6.70	(2 4 0)	8.78	8.70	(2 2 0)	12.3	12.2	(4 2 0)	10.8	10.9	(1 3 0)			
7				4.96	4.99	(6 0 0)	6.98	6.91	(2 3 0)	9.82	9.78	(7 1 0)	9.82	9.87	(7 1 0)			
8				ca.4.8	–	#	ca.4.8	–	#	8.75	8.75	(5 3 0)	8.75	8.75	(5 3 0)			
9				4.75	4.74	(6 2 0)				7.49	7.53	(4 4 0)	7.43	7.50	(4 4 0)			
10				3.82	–	h: (0 0 1)				ca.4.8	–	#	ca.4.8	–	#			
11																		
12																		
13																		
14																		
15																		

(*Continued*)

Table 13. (*Continued*)

Entry No.	○ (25)Azumane16			○ (26)Azumane17			(27)Nakai4			(28)Azumane18			(29)Ito1			(30)Ito2		
Reference	p74, 4c, $[(C_{12}O)_2Ph]_8PzNi = (C_{12}O)_{16}$-Ni			p74, 3b, $[(C_{10}O)_2Ph]_8PzCu = (C_{10}O)_{16}$-Cu			p92: $[(C_{14}O)_4BPP]_2Ce$: 14(d)			p74, 3b, $[(C_{10}O)_2Ph]_8PzCu = (C_{10}O)_{16}$-Cu			p182, $C_8PcSiCl_2$: 1a			p182, $C_8PcSiCl_2$ (Polymer): 1a		
Mw	Mw=4137.11			Mw=3693.09			Mw=2756.16			Mw=3693.09			Mw=1637.24			Mw=1637.24		
Temperature	200°C			150°C			95°C			245°C			RT			250°C		
Mesophase	Col_{rd}(C2/m)			Col_{rd}(C2/m)			$D_{LC}(=D_{L2})$			Col_{rd}(C2/m)			Col_{ro} ($P2_1/a$)			Col_{ho}		
Lattice constants	a = 70.8, b = 32.6			a = 69.7, b = 29.9			c = 42.1, h_2 = ca. 8.6, h_1 = ca.3.7			a = 68.8, b = 30.0			a = 39.5, b = 31.0, h_1 = ca. 3.9, Z = 2.0 for ρ = 1.2			a = 30.0, h_2 = ca. 6.6, Z = 2.0 for ρ = 1.05		
Peak No.	$d_{obs.}$	$d_{calcd.}$	*(hkl)*	$d_{obs.}$	$d_{calcd.}$	*(hkl)*	$d_{obs.}$	$d_{calcd.}$	*(hkl)*	$d_{obs.}$	$d_{calcd.}$	*(hkl)*	$d_{obs.}$	$d_{calcd.}$	*(hkl)*	$d_{obs.}$	$d_{calcd.}$	*(hkl)*
1	35.4	35.4	(2 0 0)	34.8	34.8	(2 0 0)	41.4	42.1	(0 0 1)	34.4	34.4	(2 0 0)	24.4	24.4	(1 1 0)	26.0	26.0	(1 0 0)
2	29.6	29.6	(1 1 0)	27.5	27.5	(1 1 0)	21.1	21.0	(0 0 2)	27.5	27.5	(1 1 0)	19.8	19.8	(2 0 0)	15.0	15.0	(1 1 0)
3	19.3	19.1	(3 1 0)	18.2	18.3	(3 1 0)	14.2	14.0	(0 0 3)	17.5	17.2	(4 0 0)	14.6	14.4	(1 2 0)	13.0	13.1	(2 0 0)
4	18.1	17.7	(4 0 0)	15.0	15	(0 2 0)	ca. 8.6	–	h_2	14.9	15	(0 2 0)	10.2	10.3	(3 2 0)	9.84	9.84	(2 1 0)
5	16.3	16.3	(0 2 0)	13.8	13.7	(2 2 0)	ca. 4.4	–	#	13.8	13.7	(2 2 0)	6.94	7.04	(5 2 0)	ca. 6.6	–	h_2: (0 0 1)
6	ca.4.8	–	#	11.4	11.3	(4 2 0)	ca. 3.7	–	h_1	11.4	11.5	(6 0 0)	ca. 4.4	–	#	ca. 4.3	–	#
7				9.16	9.16	(3 3 0)				9.90	9.90	(1 3 0)	ca. 3.9	broad	h_1: (0 0 1)			
8				8.07	8.11	(5 3 0)				9.26	9.34	(7 1 0)						
9				7.01	7.04	(7 3 0)				8.15	8.09	(5 3 0)						
10				ca.4.8	–	#				7.06	7.01	(7 3 0)						
11										ca.4.8	–	#						
12																		
13																		
14																		
15																		

#: halo of molten alkyl chains

(*Continued*)

Table 13. *(Continued)*

Entry No.	(31)Ito3			(32)Ito4			(33)Ito5			(34)Ito6			(35)Nishizawa1			(36)Nishizawa2		
Reference	p182, $C_8PcSi_2(OH)_4$ (Polymer): 3a			p182, $C_{12}Si(OH)_4$ (Polymer): 3b			p182, Doubledecker–$C_{12}Pc_2Si_2Cl_2$: 4b			p182, Doubledecker–$C_{12}Pc_2Si_2Cl_2$: 4b			Bachelor Thesis, $(C_8S)_8PcCu$: 3a			p78, 2c: $(C_{12}S)_8PcCu$ at 125°C, Col_h		
Mw	Mw=1662.45			Mw=2111.31			Mw=4101.29			Mw=4101.29			Mw=1730.33			Mw=2179.19		
Temperature	250°C			250°C			RT			140°C			125°C			125°C		
Mesophase	Col_{ho}			Col_{ho}			Col_{ho}			Col_{ho}			Col_{ho}			Col_{ho}		
Lattice constants	a = 29.3, h_2 = ca. 6.4; h_1 = ca.3.5, Z = 1.0 for ρ = 1.05			a = 33.7, h_2 = ca. 6.5, h_1 = ca.3.5, Z = 0.98 for ρ = 1.0			a = 34.5, h_2 = ca. 7.1, Z = 1.0 for ρ = 0.95			a = 35.7, h_2 = ca. 7.1, Z = 1.0 for ρ = 0.90			a = 27.5, h_2 = ca. 6.7, h_1 = ca.3.5, Z = 0.98 for ρ = 1.2			a = 32.2, h_2 = ca. 6.5, h_1 = ca.3.6, Z = 0.98 for ρ = 1.1		
Peak No.	$d_{obs.}$	$d_{calcd.}$	*(hkl)*	$d_{obs.}$	$d_{calcd.}$	*(hkl)*	$d_{obs.}$	$d_{calcd.}$	*(hkl)*	$d_{obs.}$	$d_{calcd.}$	*(hkl)*	$d_{obs.}$	$d_{calcd.}$	*(hkl)*	$d_{obs.}$	$d_{calcd.}$	*(hkl)*
1	25.1	25.4	(1 0 0)	29.0	29.2	(1 0 0)	29.8	29.8	(1 0 0)	31.2	30.9	(1 0 0)	23.8	24.0	(1 0 0)	27.9	29.7	(1 0 0)
2	14.7	14.7	(1 1 0)	16.9	16.8	(1 1 0)	17.2	17.2	(1 1 0)	17.7	17.9	(1 1 0)	13.9	13.9	(1 1 0)	16.2	16.1	(1 1 0)
3	12.7	12.7	(2 0 0)	14.6	14.6	(2 0 0)	14.9	14.9	(2 0 0)	15.7	15.5	(2 0 0)	11.9	12.0	(2 0 0)	14.0	13.9	(2 0 0)
4	9.64	9.59	(2 1 0)	11.2	11.0	(2 1 0)	ca. 7.1	–	h_2: (0 0 1)	11.5	11.7	(2 1 0)	9.20	9.08	(2 1 0)	10.6	10.5	(2 1 0)
5	ca. 6.4	–	h_2: (0 0 1)	ca. 6.5	–	h_2: (0 0 1)	ca. 4.5	–	#	ca. 7.1	–	h_2: (0 0 1)	ca. 6.7	–	h_2: (0 0 1)	ca. 6.5	–	h_2: (0 0 1)
6	ca. 4.4	–	#	ca. 4.6	–	#	ca. 3.4	–	h_1: (0 0 1)	ca. 4.6	–	#	ca.4.6	–	#	ca. 4.6	–	#
7	ca. 3.5	–	h_1: (0 0 1)	ca. 3.5	–	h_1: (0 0 1)							ca. 3.6	–	h_1: (0 0 1)	ca. 3.6	–	h_1: (0 0 1)
8																		
9	h_2: stacking distance between the dimers												h_2: stacking distance between the dimers					
10	h_1: stacking distance between the monomers												h_1: stacking distance between the monomers					
11																		
12																		
13																		
14																		
15																		

(Continued)

Table 13. *(Continued)*

Entry No.	(37)Nishizawa3			(38)Aoki1			(39)Nakai1			○ (40)Nakai2			◎ (41)Nakai3			(42)Nakai5		
Reference	Bachelor Thesis, $[(C_8S)_8Pc]_2Lu$: 4a			Bachelor Thesis, $(C_{12}O)_{15}TzCu$			p92, $(C_{10}O)_4BPPH_2$: 10(b)			p92, $(C_{12}O)_4BPPH_2$: 12(c)			p92, $(C_{14}O)_4BPPH_2$: 14(d)			p92, $[(C_8O)_4TPP]_2Ce$: 8(a)		
Mw	Mw=3508.53			Mw=3[illegible]41.86			Mw=1087.63			Mw=1199.85			Mw=1312.06			Mw=3417.01		
Temperature	125°C			RT			150°C			125°C			150°C			RT		
Mesophase	Col_{ho}			Col_{ro}[illegible]2m)			D_{L2} (= Col_L)			D_{L2} (= Col_L)			D_{L2} (= Col_L)			$Col_{rd}(P2_1/a)$		
Lattice constants	a = 26.4, h_2 = ca. 6.2, h_1 = ca.4.0 (Z = 0.99 for h_2 =8.0 ρ = 1.2)			a = 42.2, b = 31.8, Z = 0.98 for h_1 =4.0 ρ = 1.2			c = 27.2, h_2 = ca. 6.5			c = 30.6, h_2 = ca. 6.5			c = 33.5, h_2 = ca. 6.5			a = 41.4, b = 40.2		
Peak No.	$d_{obs.}$	$d_{calcd.}$	(*hkl*)	$d_{obs.}$	$d_{calcd.}$	(*hkl*)	$d_{obs.}$	$d_{calcd.}$	(*hkl*)	$d_{obs.}$	$d_{calcd.}$	(*hkl*)	$d_{obs.}$	$d_{calcd.}$	(*hkl*)	$d_{obs.}$	$d_{calcd.}$	(*hkl*)
1	22.7	22.9	(1 0 0)	31.8	31.8	(0 1 0)	26.5	27.2	(0 0 1)	30.9	30.9	(0 0 1)	33.1	33.5	(0 0 1)	28.8	28.8	(1 1 0)
2	13.2	13.2	(1 1 0)	25.4	25.4	(1 1 0)	13.7	13.6	(0 0 2)	15.2	15.3	(0 0 2)	16.9	16.8	(0 0 2)	18.1	18.1	(1 2 0)
3	11.4	11.4	(2 0 0)	17.4	17.6	(2 1 0)	9.21	9.07	(0 0 3)	10.2	10.2	(0 0 3)	11.2	11.2	(0 0 3)	13.1	13.1	(3 1 0)
4	8.64	8.53	(2 1 0)	15.2	14.9	(1 2 0)	ca. 6.5	–	h_2	ca. 6.5	–	h_2	ca. 6.5	–	h_2	10.4	10.4	(4 0 0)
5	ca. 6.2	–	h_2: (0 0 1)	12.9	12.9	(3 1 0)	ca. 4.8	–	#	ca. 4.8	–	#	ca. 4.7	–	#	9.32	9.21	(4 2 0)
6	ca. 4.7	–	#	8.57	8.46	(3 3 0)										7.95	7.89	(1 5 0)
7	ca. 4.0	–	h_1: (0 0 1)	7.23	7.43	(2 4 0)										6.92	6.91	(6 0 0)
8				ca. 4.4	–	#										6.17	6.14	(6 3 0)
9				ca. 4.0	–	h_1: (0 0 1)										ca. 4.2	–	#
10																		
11																		
12																		
13																		
14																		
15																		

(Continued)

Table 13. *(Continued)*

Entry No.	(43)Nakai6			(44)Hatada1			(45)Hatada2			(46)Hatada3			(47)Hatada4			(48)Ban1		
Reference	p92, $[(C_{10}O)_4TPP]_2Ce$: 10(b)			Master Thesis, C_9O-Ni			Master Thesis, $C_{10}O$-Ni			Master Thesis, $C_{11}O$-Ni			Master Thesis, $C_{12}O$-Ni			p72, $[Ce_2\{(C_{10}O)_4bpp\}_3]$:1a		
Mw	Mw=3865.87			Mw=1138.67			Mw=1194.77			Mw=1250.88			Mw=1306.999			Mw=3537.09		
Temperature	RT			RT			RT			RT			RT			50°C		
Mesophase	$Col_{rd}(P2_1/a)$			supercooled D_{L1}			supercooled D_{L1}			supercooled D_{L1}			supercooled D_{L1}			$D_{LC} = D_{L2}$ (= Col_L)		
Lattice constants	a = 45.4, b = 41.0			c = 26.4			c = 27.8			c = 29.1			c = 32.3			a = 40.1, h_3 = 10.4, h_2 = ca. 7.9, h_1 = ca. 3.6		
Peak No.	$d_{obs.}$	$d_{calcd.}$	*(hkl)*	$d_{obs.}$	$d_{calcd.}$	*(hkl)*	$d_{obs.}$	$d_{calcd.}$	*(hkl)*	$d_{obs.}$	$d_{calcd.}$	*(hkl)*	$d_{obs.}$	$d_{calcd.}$	*(hkl)*	$d_{obs.}$	$d_{calcd.}$	*(hkl)*
1	31.1	30.5	(1 1 0)	26.8	26.4	(0 0 1)	28.0	27.8	(0 0 1)	29.5	29.1	(0 0 1)	32.7	32.3	(0 0 1)	38.5	40.1	(0 0 1)
2	23.2	22.7	(2 0 0)	13.0	13.2	(0 0 2)	13.8	13.9	(0 0 2)	14.5	14.6	(0 0 2)	16.1	16.2	(0 0 2)	20.3	20.0	(0 0 2)
3	18.7	18.7	(1 2 0)	ca.4.4	–	#	ca.4.3	–	#	9.60	9.70	(0 0 3)	10.7	10.8	(0 0 3)	13.3	13.4	(0 0 3)
4	14.2	14.2	(3 1 0)							ca.4.3	–	#	ca.4.4	–	#	10.4	10.1	(0 0 4)+h_3
5	13.1	13.1	(1 3 0)													ca. 7.9	–	h_2
6	11.3	11.3	(4 0 0)													ca. 4.2	–	#
7	10.6	10.9	(4 1 0)													ca. 3.6	–	h_1
8	8.84	8.86	(5 1 0)															
9	8.07	8.08	(1 5 0)															
10	6.40	6.41	(7 1 0)															
11	ca. 4.2	–	#															
12																		
13																		
14																		
15																		

h_3: stacking distance between the trimers

h_2: stacking distance between the dimers

h_1: stacking distance between the monomers

(Continued)

Table 13. *(Continued)*

Entry No.	(49)Ban2			○ (50)Ban3		
Reference	p72, [Ce$_2${(C$_{10}$O)$_4$bpp}$_3$]:1a			p72, [Ce$_2${(C$_{12}$O)$_4$bpp}$_3$]:1b		
Mw	Mw=3537.09			Mw=3873.73		
Temperature	130°C			RT		
Mesophase	Col$_{ro}$(P2/a)			D$_{L2}$ (= Col$_L$)		
Lattice constants	a = 59.0, b = 33.6, h_3 = 10.7, h_2 = ca. 7.2, h_1 = ca. 3.7, Z = 4.0 for ρ = 1.1			c = 46.0, h_3 = 10.7, h_2 = ca. 6.6, h_1 = 3.60,		
Peak No.	d$_{obs.}$	d$_{calcd.}$	(*hkl*)	d$_{obs.}$	d$_{calcd.}$	(*hkl*)
1	33.6	33.6	(0 1 0)	45.8	46.0	(0 0 1)
2	29.5	29.5	(2 0 0)	22.8	22.0	(0 0 2)
3	22.3	22.2	(2 1 0)	15.4	15.4	(0 0 3)
4	17.2	17.0	(3 1 0)	10.7	10.7	h_3
5	11.5	11.2	(0 3 0)	9.25	9.48	(0 0 4)
6	10.7	10.7	h_3	ca. 6.6	–	h_2
7	8.75	8.92	(4 3 0)	ca. 4.2	–	#
8	ca. 7.2	–	h_2	3.50	–	h_1
9	ca. 4.2	–	#			
10	ca. 3.7	–	h_1			
11						
12						
13						
14						
15						

Entry No.	(51)Ban8			(52)Ban9		
Reference	p72, [Ce$_2${(C$_{12}$O)$_8$sbtp}$_3$]:2b			p72, [Ce$_2${(C$_{16}$O)$_8$sbtp}$_3$]:2d		
Mw	Mw=6998.80			Mw=8345.39		
Temperature	140°C			120°C		
Mesophase	Col$_{ho}$			Col$_{ho}$		
Lattice constants	a = 38.4, *h* = ca. 9.0, Z = 0.99 for ρ = 1.0			a = 44.8, h = ca. 9.2, Z = 1.0 for ρ = 0.90		
Peak No.	d$_{obs.}$	d$_{calcd.}$	(*hkl*)	d$_{obs.}$	d$_{calcd.}$	(*hkl*)
1	33.2	33.2	(1 0 0)	38.8	38.8	(1 0 0)
2	20.0	19.2	(1 1 0)	22.2	22.4	(1 1 0)
3	ca. 9.0	–	*h*	ca. 9.2	–	*h*
4	ca. 4.6	–	#	4.16	–	#

Entry No.	(53)Ban7			○ (54)Ban13		
Reference	p72, [Ce$_2${(C$_{12}$O)$_8$sbtp}$_3$]:2b			p78, (C$_{10}$S)$_8$PcCu: 2b		
Mw	Mw=6998.80			Mw=1954.76		
Temperature	RT			125°C		
Mesophase	glassy Col$_{ho}$			Col$_{ho}$		
Lattice constants	a = 43.4, h = ca. 9.4, Z = 1.0 for ρ = 0.85			a = 29.9, h_2 = ca. 7.2, h_1 = ca. 3.5, Z = 1.0 for ρ = 1.2		
Peak No.	d$_{obs.}$	d$_{calcd.}$	(*hkl*)	d$_{obs.}$	d$_{calcd.}$	(*hkl*)
1	35.6	35.6	(1 0 0)	26.3	25.9	(1 0 0)
2	20.4	20.6	(1 1 0)	15.0	15.0	(1 1 0)
3	ca. 9.4	–	*h*	13.0	13.0	(2 0 0)
4	ca. 4.4	–	#	9.86	9.80	(2 1 0)
5				ca. 7.2	–	h_2
6				ca. 4.7	–	#
7				ca. 3.5	–	h_1

(Continued)

Table 13. *(Continued)*

Entry No.	⊙ (55)Ban4			(56)Ban5			O (57)Ban6			(58)Ban9b			(59)Ban10			(60)Ban11		
Reference	p72, [Ce_2{($(C_{12}O)_4$bpp}$_3$]:1b			p72, [Ce_2{($(C_{14}O)_4$bpp}$_3$]:1c			p72, [Ce_2{($(C_{14}O)_4$bpp}$_3$]:1c			p72, [Ce_2{($(C_{16}O)_8$btp}$_3$]:2d			Master Thesis Chap 3, p78, 1c:$(C_{12}S)_8PcH_2$			Master Thesis Chap 3, p78, 1d:$(C_{16}S)_8PcH_2$		
Mw	Mw=3873.73			Mw=4210.35			Mw=4210.35			Mw=8345.39			Mw=2117.66			Mw=2566.52		
Temperature	140°C			50°C			125°C			120°C			125°C			125°C		
Mesophase	Col_r (P2/a)			D_{L2} (= Col_L)			Col_r (P2/a)			Col_h			Col_h			Col_h		
Lattice constants	a = 60.8, b = 38.8, h_3 = 10.9, h_2 = 7.00, h_1 = ca. 3.6, Z = 4.0 for ρ = 1.0			c = 50.4, h_3 = 10.2, h_2 = 7.38, h_1 = 3.54			a = 61.2, b = 41.1, h_3 = 10.6, h_2 = ca. 7.1, h_1 = ca. 3.6, Z = 4.0 for ρ = 1.05			a = 43.4, h = ca. 9.6, Z = 1.0 for ρ = 0.98			a = 32.4, h_2 = ca. 6.3, h_1 = ca. 3.6, Z = 1.0 for ρ = 1.1			a = 36.1, h_2 = ca. 6.4, h_1 = ca. 3.5, Z = 1.0 for ρ = 1.1		
Peak No.	$d_{obs.}$	$d_{calcd.}$	*(hkl)*	$d_{obs.}$	$d_{calcd.}$	*(hkl)*	$d_{obs.}$	$d_{calcd.}$	*(hkl)*	$d_{obs.}$	$d_{calcd.}$	*(hkl)*	$d_{obs.}$	$d_{calcd.}$	*(hkl)*	$d_{obs.}$	$d_{calcd.}$	*(hkl)*
1	38.2	38.8	(0 1 0)	24.9	25.4	(0 0 2)	40.0	41.1	(0 1 0)	37.6	37.6	(1 0 0)	28.1	28.1	(1 0 0)	31.2	31.2	(1 0 0)
2	30.4	30.4	(2 0 0)	17.0	17.0	(0 03)	30.6	30.6	(2 0 0)	21.9	21.7	(1 1 0)	16.2	16.2	(1 1 0)	18.1	18.1	(1 1 0)
3	19.4	19.4	(0 2 0)	12.7	12.7	(0 0 4)	20.5	20.5	(0 2 0)	ca. 9.6	–	*h*	ca. 6.3	–	h_2	15.6	15.6	(2 0 0)
4	16.4	16.4	(2 2 0)	10.2	10.2	(0 0 5)+h_3	13.9	13.7	(0 3 0)	ca. 4.6	–	#	ca. 4.6	–	#	ca. 6.4	–	h_2
5	15.1	15.2	(4 0 0)	8.62	8.48	(0 0 6)	10.6	10.5	(2 5 0)+h_3				ca. 3.6	–	h_1	ca. 4.7	–	#
6	14.2	14.2	(4 1 0)	7.38	7.27	(0 0 7)+h_2	ca. 7.1	–	h_2							ca. 3.5	–	h_1
7	12.9	12.9	(0 3 0)	6.31	6.35	(0 0 8)	ca. 4.3	–	#									
8	10.9	10.9	(3 3 0)+h_3	4.23	–	#2	ca. 3.6	–	h_1									
9	10.1	10.1	(6 0 0)	3.54	–	h_1												
10	8.95	8.99	(6 2 0)															
11	8.42	8.48	(7 1 0)															
12	7.76	7.76	(0 5 0)															
13	7.00	7.01	(6 4 0)+h_2															
14	ca. 4.3	–	#															
15	ca. 3.6	–	h_1															

#2: Not broad but fairly sharp.

h_3: stacking distance between the trimers

h_2: stacking distance between the dimers

h_1: stacking distance between the monomers

(Continued)

Table 13. (*Continued*)

Entry No.	○ (61)Ban12			○ (62)Hatsusaka8			○ (63)Ban14			◎(64)Ban15			(65)Ban16			○ (66)Ban30		
Reference	p78, $(C_8S)_8PcCu$: 2a			p84, $[(C_{12}O)_2PhO]_8PcCu$: 8d			P78, $(C_{12}S)_8PcCu$: 2c			p78, $(C_{16}S)_8PcCu$: 2d			p83, $[(C_8S)_8Pc]_2Lu$: 3a			p83, $[(C_{16}S)_8Pc]_2Eu$: 1e		
Mw	Mw=1730.33			Mw=4[illegible]62.01			Mw=2179.19			Mw=2628.05			Mw=3508.53			Mw=4832.11		
Temperature	125°C			165°C			125°C			125°C			125°C			120°C		
Mesophase	Col_{ho}			$Col_{rect.d}$			Col_{ho}			Col_{ho}			Col_{ho}			Col_{ho}		
Lattice constants	a = 28.0, h_2 = 6.70, h_1 = ca. 3.5; Z = 0.99 for ρ = 1.2			a = [illegible]3.4			a = 32.2, h_2 = ca. 6.5, h_1 = ca. 3.6、Z = 0.98 for ρ = 1.1			a = 35.8, h_2 = 7.09, h_1 = ca. 3.5; Z = 0. 98 for ρ = 1.1			a = 26.4, h_2 = ca. 6.2, h_1 = ca. 3.6, Z = 0.50 for h_1 and ρ = 1.35; Z = 1.0 for $2h_1$ and ρ = 1.35			a = 34.3, h_2 = ca. 7.7, Z = 0.98 for ρ = 1.0		
Peak No.	$d_{obs.}$	$d_{calcd.}$	(*hkl*)	$d_{obs.}$	$d_{calcd.}$	(*hkl*)	$d_{obs.}$	$d_{calcd.}$	(*hkl*)	$d_{obs.}$	$d_{calcd.}$	(*hkl*)	$d_{obs.}$	$d_{calcd.}$	(*hkl*)	$d_{obs.}$	$d_{calcd.}$	(*hkl*)
1	24.3	24.0	(1 0 0)	3[illegible].0	32.0	(1 0 0)	27.9	27.9	(1 0 0)	31.0	31.0	(1 0 0)	22.7	22.9	(1 0 0)	29.7	29.7	(1 0 0)
2	14.0	13.9	(1 1 0)	2[illegible].8	21.8	(1 1 0)	16.2	16.1	(1 1 0)	18.0	17.9	(1 1 0)	13.2	13.2	(1 1 0)	16.9	17.2	(1 1 0)
3	12.1	12.0	(2 0 0)	1[illegible].2	12.1	(3 2 0)	14.0	13.9	(2 0 0)	15.6	15.6	(2 0 0)	11.4	11.4	(2 0 0)	15.2	14.9	(2 0 0)
4	9.21	9.08	(2 1 0)	ca. 4.6	–	#	10.6	10.5	(2 1 0)	11.9	11.7	(2 1 0)	8.64	8.53	(2 1 0)	10.2	9.90	(3 0 0)
5	8.20	8.09	(3 0 0)				ca. 6.5	–	h_2	8.98	8.96	(2 2 0)	ca. 6.2	–	h_2	ca.7.7	–	h_2
6	6.70	6.73	(1 3 0)+h_2				ca. 4.6	–	#	7.09	7.12	(2 3 0)+h_2	ca. 4.7	–	#	ca.4.6	–	#
7	ca. 4.6	–	#				ca. 3.6	–	h_1	5.92	5.97	(3 3 0)	ca. 3.6	–	h_1			
8	ca. 3.5	–	h_1							5.16	5.17	(0 6 0)						
9										ca. 4.6	–	#						
10										ca. 3.5	–	h_1						
11																		
12																		
13																		
14																		
15																		

(*Continued*)

Table 13. *(Continued)*

Entry No.	○ (67)Ban17					○ (68)Ban18			(69)Ban19					○ (70)Ban20		
Reference	p83, [$(C_{10}S)_8Pc]_2Lu$: 3b					p83, [$(C_{10}S)_8Pc]_2Lu$: 3b			p83, [$(C_{12}S)_8Pc]_2Lu$: 3c					p83, [$(C_{12}S)_8Pc]_2Lu$: 3c		
Mw	Mw=3957.39					Mw=3957.39			Mw=4406.25					Mw=4406.25		
Temperature	RT					120°C			32°C					120°C		
Mesophase	pseudo-Col_{ho}					Col_{ho}			pseudo-Col_{ho}					Col_{ho}		
Lattice constants	Hex: a = 31.3, h_2 = ca.7.6, h_1 = 3.28, Z = 0.98 for h_2 and ρ = 1.0; Rec: a = 54.2, b = 31.3 = hex a, h_2 = ca.7.6, h_1 = 3.28, Z = 2.0 for h_2 and ρ = 1.0					a = 29.0, h_2 = ca.6.7, Z = 1.0 for ρ = 1.4			Hex: a = 33.8, h_2 = ca.6.5, h_1 = 3.28, Z = 1.0 for h2 and ρ = 1.15; Rec: a = 58.4, b = 33.8 = hex a, h_2 = ca.6.5, h_1 = 3.28 ; Z = 2.0 for h_2 and ρ = 1.15					a = 30.7, h_2 = ca.7.0, Z = 1.0 for ρ = 1.3		
Peak No.	$d_{obs.}$	Hex $d_{calcd.}$	Rec $d_{calcd.}$	Hex *(hkl)*	Rec *(hkl)*	$d_{obs.}$	$d_{calcd.}$	*(hkl)*	$d_{obs.}$	Hex $d_{calcd.}$	Rec $d_{calcd.}$	Hex *(hkl)*	Rec *(hkl)*	$d_{obs.}$	$d_{calcd.}$	*(hkl)*
1	26.6	27.1	27.1	(1 0 0)	(1 0 0), (1 1 0)	25.1	25.1	(1 0 0)	29.1	29.2	29.1	(1 0 0)	(1 0 0), (1 1 0)	26.6	26.6	(1 0 0)
2	21.0	–	20.5	–	(2 1 0)	14.5	14.5	(1 1 0)	22.5	–	22.1	–	(2 1 0)	15.5	15.5	(1 1 0)
3	15.6	15.6	15.6	(1 1 0)	(0 2 0), (3 1 0)	12.6	12.5	(2 0 0)	16.8	16.9	16.9	(1 1 0)	(0 2 0), (3 1 0)	13.4	13.4	(2 0 0)
4	13.6	13.5	13.6	(2 0 0)	(2 2 0), (4 0 0)	9.51	9.48	(2 1 0)	14.7	14.6	14.6	(2 0 0)	(2 2 0), (4 0 0)	10.2	10.2	(2 1 0)
5	10.3	10.2	10.2	(2 1 0)	(4 2 0), (5 1 0)	8.47	8.36	(3 0 0)	11.2	11.0	11.3	(2 1 0)	(4 2 0), (5 1 0)	7.73	7.73	(2 2 0)
6	ca.7.6	–	–	h_2	h_2	ca.6.7	–	h_2	9.88	9.74	9.74	(3 0 0)	(3 3 0), (6 0 0)	ca.7.0	–	h_2
7	ca.4.3	–	–	#	#	ca.4.7	–	#	8.26	8.10	8.35	(3 1 0)	(1 4 0)	ca.4.7	–	#
8	3.28	–	–	h_1	h_1				7.42	7.31	7.37	(4 0 0)	(6 3 0)			
9									ca.6.5	–	–	h_2	h_2			
10									ca.4.4	–	–	#	#			
11									3.28	–	–	h_1	h_1			
12																
13																
14																
15																

(Continued)

Table 13. *(Continued)*

Entry No.	○ (71)Ban21			(72)Ban22			(73)Ban23			◉ (74)Ban24			○ (75)Ban25				
Reference	p83, $[(C_{14}S)_8Pc]_2Lu$: 3d			p83, $[(C_{16}S)_8Pc]_2Lu$: 3e			p83, $[(C_{14}S)_8Pc]_2Lu$: 3f			p83, $[(C_8S)_8Pc]_2Eu$: 1a			p83, $[(C_{10}S)_8Pc]_2Eu$: 1b				
Mw	Mw=4855.11			Mw=5303.97			Mw=5752.83			Mw=3485.53			Mw=3934.39				
Temperature	120°C			120°C			100°C			120°C			50°C				
Mesophase	Col_{ho}			Col_{ho}			Col_{ho}			Col_{ho}			pseudo-Col_{ho}				
Lattice constants	a = 32.9, h_2 = ca.7.8; Z = 1.0 for ρ = 1.1			a = 35.1, h_2 = ca.7.18; Z = 1.0 for ρ = 1.2			a = 38.3, h_2 = ca.7.08; Z = 1.0 for ρ = 1.1			a = 27.9, h_2 = ca.7.3, h_1 = ca.3.6; Z = 1.0 for h_2 and ρ = 1.2			Hex: a = 31.9, h_2 = 6.90, h_1 = ca. 3.3, Z = 1.0 for ρ = 1.1; Rec: a = 55.2, b = 31.9 = hex a, h_2 = 7.43, h_1 = ca. 3.3, Z = 2.0 for ρ = 1.1				
Peak No.	$d_{obs.}$	$d_{calcd.}$	*(hkl)*	$d_{obs.}$	$d_{calcd.}$	*(hkl)*	$d_{obs.}$	$d_{calcd.}$	*(hkl)*	$d_{obs.}$	$d_{calcd.}$	*(hkl)*	$d_{obs.}$	Hex $d_{calcd.}$	Rec $d_{calcd.}$	Hex *(hkl)*	Rec *(hkl)*
1	27.8	28.5	(1 0 0)	30.4	30.4	(1 0 0)	33.5	33.2	(1 0 0)	24.5	24.2	(1 0 0)	27.2	27.6	27.6	(1 0 0)	(1 0 0), (1 1 0)
2	16.5	16.5	(1 1 0)	17.8	17.8	(1 1 0)	19.2	19.1	(1 1 0)	14.0	14.0	(1 1 0)	20.8	–	20.9	–	(2 1 0)
3	14.2	14.3	(2 0 0)	15.4	15.4	(2 0 0)	16.5	16.6	(2 0 0)	12.1	12.1	(2 0 0)	15.9	15.9	15.9	(1 1 0)	(0 2 0), (3 1 0)
4	10.8	10.8	(2 1 0)	ca.7.1	–	h_2	12.3	12.5	(2 1 0)	9.17	9.14	(2 1 0)	13.7	13.8	13.8	(2 0 0)	(2 2 0), (4 0 0)
5	9.52	9.50	(3 0 0)	ca.4.7	–	#	ca.7.0	–	h_2	7.92	8.06	(3 0 0)	12.2	–	12.0	–	(3 2 0)
6	ca. 7.8	–	h_2				ca.4.7	–	#	ca. 7.3	–	h_2	10.5	10.4	10.4	(2 1 0)	(4 2 0), (5 1 0)
7	ca. 4.7	–	#							6.70	6.71	(3 1 0)	9.21	9.20	9.20	(3 0 0)	(3 3 0), (6 0 0)
8										ca. 4.7	–	#	8.04	7.96	7.96	(2 2 0)	(0 4 0), (6 2 0)
9										ca. 3.6	–	h_1	7.74	7.65	7.74	(3 1 0)	(2 4 0), (5 3 0)
10													6.90	6.90	6.90	(4 0 0)+h_2	(4 4 0)+h_2
11													ca. 4.6	–	–	#	#
12													4.28	–	4.32	–	(4 7 0)
13													ca. 3.3	–	–	h_1	h_1
14																	
15																	

(Continued)

Table 13. *(Continued)*

Entry No.	(76)Ban26			⊙ (77)Ban27					⊙ (78)Ban28			(79)Ban29			(80)Ban31		
Reference	p83, $[(C_{10}S)_8Pc]_2Eu$: 1b			p83, $[(C_{10}S)_8Pc]_2Eu$: 1c					p83, $[(C_{12}S)_8Pc]_2Eu$: 1c			p83, $[(C_{14}S)_8Pc]_2Eu$: 1d			p83, $[(C_8S)_8Pc]_2Tb$: 2a		
Mw	Mw=3934.39			Mw=4383.25					Mw=4383.25			Mw=4832.11			Mw=3492.49		
Temperature	120°C			45°C					120°C			120°C			120°C		
Mesophase	Col$_h$			pseudo-Col$_h$					Col$_h$			Col$_h$			Col$_h$		
Lattice constants	a = 29.9, h_2 = 7.23; Z = 1.0 for ρ = 1.2			Hex: a = 34.0, h_2 = 7.43, h_1 = 3.30, Z = 1.0 for ρ= 1.0; Rec: a = 58.9, b = 34.0 = hex a, h_2 = 7.43, h_1 = 3.30, Z = 2.0 for ρ= 1.0					a = 31.8, h_2 = ca.7.2; Z = 1.0 for ρ= 1.20			a = 34.2, h_2 = ca.7.6, h_1 = ca.3.5; Z = 1.0 for ρ= 1.05			a = 27.9, h_2 = ca.7.4; Z = 1.0 for ρ= 1.20		
Peak No.	$d_{obs.}$	$d_{calcd.}$	*(hkl)*	$d_{obs.}$	Hex $d_{calcd.}$	Rec $d_{calcd.}$	Hex *(hkl)*	Rec *(hkl)*	$d_{obs.}$	$d_{calcd.}$	*(hkl)*	$d_{obs.}$	$d_{calcd.}$	*(hkl)*	$d_{obs.}$	$d_{calcd.}$	*(hkl)*
1	25.5	25.9	(1 0 0)	28.8	29.5	28.8	(1 0 0)	(1 1 0), (2 0 0)	27.0	27.5	(1 0 0)	29.8	29.6	(1 0 0)	24.1	24.1	(1 0 0)
2	14.9	14.9	(1 1 0)	22.0	–	22.3	–	(2 1 0)	15.9	15.9	(1 1 0)	17.1	17.1	(1 1 0)	14.0	14.0	(1 1 0)
3	12.9	12.9	(2 0 0)	17.0	17.0	17.0	(1 1 0)	(0 2 0), (3 1 0)	13.7	13.8	(2 0 0)	15.0	14.8	(2 0 0)	12.4	12.1	(2 0 0)
4	9.82	9.79	(2 1 0)	14.8	14.7	14.7	(2 0 0)	(2 2 0), (4 0 0)	10.5	10.4	(2 1 0)	11.1	11.2	(2 1 0)	9.07	9.11	(2 1 0)
5	8.67	8.63	(3 0 0)	11.2	11.1	11.1	(2 1 0)	(1 3 0), (4 2 0)	9.23	9.17	(3 0 0)	8.08	8.22	(3 0 0)	7.94	8.04	(3 0 0)
6	7.23	7.18	(3 1 0)+h_2	9.85	9.82	9.82	(3 0 0)	(3 3 0), (6 0 0)	7.52	7.63	(3 1 0)	ca.7.6	–	h_2	ca. 7.4	–	h_2
7	6.08	5.94	(3 2 0)	9.45	–	9.43	–	(6 1 0)	ca.7.2	–	h_2	5.39	5.32	(5 1 0)	6.55	6.69	(3 1 0)
8	ca.4.7	–	#	8.25	8.17	8.17	(3 1 0)	(2 4 0)	ca.4.7	–	#	ca.4.8	–	#	5.43	5.53	(2 3 0)
9				7.74	–	7.80	–	(3 4 0)				ca.3.5	–	h_1	ca. 4.7	–	#
10				7.43	7.36	7.42	(4 0 0)+h_2	(6 3 0)+h_2									
11				5.65	5.65	5.67	(3 3 0)	(0 6 0)									
12				ca. 4.5	–	–	#	#									
13				3.30	–	–	h_1	h_1									
14																	
15																	

(Continued)

Table 13. *(Continued)*

Entry No.	(81)Ban32					(82)Ban33			O (83)Ban34					(84)Ban35		
Reference	p83, [(C10S)8Pc]2Tb: 2b					p83, [(C10S)8Pc]2Tb: 2b			p83, [(C12S)8Pc]2Tb: 2c					p83, [(C12S)8Pc]2Tb: 2c		
Mw	Mw=3941.35					Mw=3941.35			Mw=4390.21					Mw=4390.21		
Temperature	40°C					120°C			40°C					120°C		
Mesophase	pseudo-Col$_{ho}$					Col$_{ho}$			pseudo-Col$_{ho}$					Col$_{ho}$		
Lattice constants	Hex: a = 31.5, h_2 = ca.6.5, h_1 = 3.27, Z = 1.0 for ρ= 1.20; Rec: a = 54.5, b = 31.5 = hex a, h_2 = ca. 6.5, h_1 = 3.27, Z = 2.0 for ρ= 1.20					a = 29.4, h_2 = 7.31; Z = 1.0 for ρ= 1.20			Hex: a = 34.1, h_2 = 6.50, h_1 = 3.29, Z = 1.0 for ρ= 1.10; Rec: a = 59.1, b = 34.1, h_2 = 6.50, h_1 = 3.29, Z = 2.0 for ρ= 1.10					a = 31.4, h_2 = ca.7.4; Z = 1.0 for ρ= 1.20		
Peak No.	$d_{obs.}$	Hex $d_{calcd.}$	Rec $d_{calcd.}$	Hex (*hkl*)	Rec (*hkl*)	$d_{obs.}$	$d_{calcd.}$	(*hkl*)	$d_{obs.}$	Hex $d_{calcd.}$	Rec $d_{calcd.}$	Hex (*hkl*)	Rec (*hkl*)	$d_{obs.}$	$d_{calcd.}$	(*hkl*)
1	27.0	27.3	27.3	(1 0 0)	(1 0 0), (1 1 0)	25.1	25.5	(1 0 0)	29.1	29.5	29.5	(1 0 0)	(1 0 0), (1 1 0)	27.2	27.2	(1 0 0)
2	20.7	–	20.6	–	(2 1 0)	14.7	14.7	(1 1 0)	22.2	–	22.3	–	(2 1 0)	15.9	15.9	(1 1 0)
3	15.7	15.7	15.7	(1 1 0)	(0 2 0), (3 1 0)	12.9	12.7	(2 0 0)	17.1	17.1	17.1	(1 1 0)	(0 2 0), (3 1 0)	13.7	13.7	(2 0 0)
4	13.7	13.6	13.6	(2 0 0)	(2 2 0), (4 0 0)	9.74	9.62	(2 1 0)	14.7	14.8	14.8	(2 0 0)	(2 2 0), (4 0 0)	9.17	9.08	(3 0 0)
5	10.4	10.3	10.3	(2 1 0)	(1 3 0)	8.48	8.48	(3 0 0)	11.3	11.2	11.2	(2 1 0)	(4 2 0)	ca.7.4	–	h_2
6	9.27	9.09	9.01	(3 0 0)	(3 3 0), (6 0 0)	7.31	7.34	(2 2 0)+h_2	9.82	9.85	9.85	(3 0 0)	(3 3 0)	5.62	5.45	(5 0 0)
7	7.62	7.56	7.56	(3 1 0)	(2 4 0)	6.42	6.36	(4 0 0)	8.28	8.20	8.19	(3 1 0)	(5 3 0), (2 4 0)	ca.4.7	–	#
8	ca. 6.5	6.45	6.45	h_2	h_2	ca.4.7	–	#	6.50	6.45	6.45	(4 1 0)+h_2	(3 5 0)+h_2	3.91	3.89	(3 5 0)
9	6.01	5.95	5.95	(4 1 0)	(3 5 0)				ca. 4.4	–	–	#	#			
10	5.53	5.46	5.54	(5 0 0)	(7 4 0)				3.29	–	–	h_1	h_1			
11	ca. 4.5	–	–	#	#											
12	3.27	–	–	h_1	h_1											
13																
14																
15																

(Continued)

Table 13. (*Continued*)

	(85)Ban36			(86)Ban37		
Reference	p83, [($C_{14}S)_8Pc]_2Tb$: 2d			p83, [($C_{16}S)_8Pc]_2Tb$: 2e		
Mw	Mw=4839.07			Mw=5287.93		
Temperature	120°C			120°C		
Mesophase	Col_{ho}			Col_{ho}		
Lattice constants	a = 33.5, h_2 = ca.7.6; Z = 1.0 for ρ= 1.10			a = 34.6, h_2 = ca.7.4; Z = 1.0 for ρ= 1.15		
Peak No.	$d_{obs.}$	$d_{calcd.}$	*(hkl)*	$d_{obs.}$	$d_{calcd.}$	*(hkl)*
1	29.0	29.0	(1 0 0)	29.9	29.9	(1 0 0)
2	16.8	16.8	(1 1 0)	17.7	17.3	(1 1 0)
3	14.3	14.5	(2 0 0)	15.3	15	(2 0 0)
4	11.0	11.0	(2 1 0)	11.7	11.3	(2 1 0)
5	9.57	9.68	(3 0 0)	8.50	8.64	(2 2 0)
6	8.13	8.05	(3 1 0)	ca.7.4	–	h_2
7	ca.7.6	–	h_2	6.81	6.81	(2 3 0)
8	ca.4.7	–	#	ca.4.7	–	#
9						
10						
11						
12						
13						
14						
15						

	◎ (87)Ban38			(88)Hatsusaka1		
Reference	p83, [($C_{18}S)_8Pc]_2Tb$: 2f			p84, [($(C_9O)_2PhO]_8PcCu$: 8a		
Mw	Mw=5736.79			Mw=3588.72		
Temperature	80°C			68°C		
Mesophase	Col_{ho}			Col_{hd}		
Lattice constants	a = 38.2, h_2 = ca.7.0, h_1 = ca.3.5; Z = 1.0 for h_2 = 7.0 and ρ= 1.1			a = 36.4		
Peak No.	$d_{obs.}$	$d_{calcd.}$	*(hkl)*	$d_{obs.}$	$d_{calcd.}$	*(hkl)*
1	33.1	33.1	(1 0 0)	31.5	31.5	(1 0 0)
2	19.2	19.1	(1 1 0)	11.9	11.9	(2 1 0)
3	16.6	16.6	(2 0 0)	8.78	8.74	(3 1 0)
4	12.7	12.5	(2 1 0)	6.91	6.88	(4 1 0)
5	9.61	9.56	(2 2 0)	ca.4.5	–	#
6	ca.7.0	–	h_2			
7	6.41	6.37	(3 3 0)			
8	ca.4.5	–	#			
9	ca.3.5	–	h_1			
10						
11						
12						
13						
14						
15						

	○ (89)Hatsusaka5			(90)Hatsusaka2		
Reference	p84, [($(C_{12}O)_2PhO]_8PcCu$: 8d			p84, [($(C_9O)_2PhO]_8PcCu$: 8a		
Mw	Mw=4262.01			Mw=3588.72		
Temperature	94°C			116°C		
Mesophase	Col_{hd}			$Col_{ro}(P2_1/a)$		
Lattice constants	a = 41.0			a = 66.2, b = 32.9, h_2 = 9.96, Z = 4.0 for ρ= 1.1; h_1 = 4.98, Z = 2.0 for ρ= 1.1		
Peak No.	$d_{obs.}$	$d_{calcd.}$	*(hkl)*	$d_{obs.}$	$d_{calcd.}$	*(hkl)*
1	35.5	35.5	(1 0 0)	33.1	33.1	(2 0 0)
2	20.5	20.5	(1 1 0)	29.4	29.4	(1 1 0)
3	13.4	13.4	(2 1 0)	24.1	22.3	(2 1 0)
4	10.2	10.2	(2 2 0)	13.5	13.2	(3 2 0)
5	ca.4.5	–	#	12.4	12.3	(5 1 0)
6				11.3	11	(6 0 0)
7				10.1	9.96	(0 0 1): h_2
8				9.07	9.08	(7 1 0)
9				8.22	8.21	(0 4 0)
10				7.87	7.77	(6 3 0)
11				6.48	6.44	(2 5 0)
12				4.98	4.98	(0 0 2): h_1
13				ca.4.5	–	#
14						
15						

(*Continued*)

Table 13. (*Continued*)

Entry No.	◎ (91)Hatsusaka3			◎ (92)Hatsusaka4		
Reference	p84, [(C$_9$O)$_2$PhO]$_8$PcCu: 8a			p84, [(C$_9$O)$_2$PhO]$_8$PcCu: 8a		
Mw	Mw=3588.72			Mw=3588.72		
Temperature	140°C			156°C		
Mesophase	Col$_{ro}$(P2$_1$/a)			Col$_{rd}$(P2m)		
Lattice constants	a = 66.2, b = 32.4, h_2 = 10.0, Z = 4.0 for ρ= 1.1; h_1 = 5.0, Z = 2.0 for ρ= 1.1			a =53.2, b =30.2, Z = 1.0 for h = 3.7[c], ρ= 1.0[c]		
Peak No.	d$_{obs.}$	d$_{calcd.}$	(*hkl*)	d$_{obs.}$	d$_{calcd.}$	(*hkl*)
1	33.1	33.1	(2 0 0)	30.2	30.2	(0 1 0)
2	29.1	29.1	(1 1 0)	26.6	26.6	(2 0 0)
3	13.2	13.0	(3 2 0)	20.0	20.0	(2 1 0)
4	11.2	11.0	(6 0 0)	11.5	11.5	(3 2 0)
5	10.2	10.2+10.0	(5 2 0)+h_2	8.52	8.51	(6 1 0)
6	9.64	9.69	(3 3 0)	ca.4.6	–	#
7	8.15	8.16	(7 2 0)			
8	7.32	7.27	(4 4 0)	c: assumed value		
9	6.40	6.44	(1 5 0)			
10	5.92	5.91	(11 1 0)			
11	ca.5.0	5.0	(0 0 2): h_1			
12	ca.4.5	–	#			
13						
14						
15						

Entry No.	(93)Hatsusaka6			◎ (94)Hatsusaka7		
Reference	p84, [(C$_{12}$O)$_2$PhO]$_8$PcCu: 8d			p84, [(C$_{12}$O)$_2$PhO]$_8$PcCu: 8d		
Mw	Mw=4262.01			Mw=4262.01		
Temperature	120°C			141°C		
Mesophase	Col$_{ro}$(P2$_1$/a)			Col$_{rd}$(P2$_1$/a)		
Lattice constants	a = 73.6, b = 36.7, h_2 = 9.96, Z = 4.0 for ρ= 1.05; h_1 = 4.98, Z = 2.0 for ρ= 1.05			a = 64.7, b = 59.7		
Peak No.	d$_{obs.}$	d$_{calcd.}$	(*hkl*)	d$_{obs.}$	d$_{calcd.}$	(*hkl*)
1	36.8	36.8	(2 0 0)	32.3	32.3	(2 0 0)
2	33.7	32.9	(1 1 0)	29.9	29.9	(0 2 0)
3	27.5	26	(2 1 0)	21.9	21.9	(2 2 0)
4	20.0	20.4	(3 1 0)	12.3	12.3	(3 4 0)
5	14.6	14.6	(3 2 0)	9.05	9.04	(3 6 0)
6	13.6	13.7	(5 1 0)	ca.4.6	–	#
7	12.9	13.0	(4 2 0)			
8	11.0	11.0	(3 3 0)			
9	9.96	9.96	(0 0 1): h_2			
10	8.86	8.91	(2 4 0)			
11	8.58	8.6	(3 4 0)			
12	7.87	7.87	(5 4 0)			
13	7.36	7.35	(6 4 0)			
14	4.98	4.98	(0 0 2): h_1			
15	ca.4.5	–	#			

Entry No.	○ (95)Hatsusaka9			(96)Hatsusaka10		
Reference	p84, [(C$_{12}$O)$_2$PhO]$_8$PcCu: 8f			p84, [(C$_{12}$O)$_2$PhO]$_8$PcCu: 8f		
Mw	Mw=4710.87			Mw=4710.87		
Temperature	98°C			122°C		
Mesophase	Col$_{hd}$			Col$_{rd}$(P2$_1$/a)		
Lattice constants	a = 43.2			a = 77.8, b = 38.8		
Peak No.	d$_{obs.}$	d$_{calcd.}$	(*hkl*)	d$_{obs.}$	d$_{calcd.}$	(*hkl*)
1	37.4	37.4	(1 0 0)	38.9	38.9	(2 0 0)
2	21.6	21.6	(1 1 0)	36.0	34.7	(1 1 0)
3	18.7	18.7	(2 0 0)	21.4	21.6	(3 1 0)
4	14.2	14.1	(2 1 0)	18.1	17.4	(4 1 0)
5	10.7	10.8	(2 2 0)	15.3	15.5	(3 2 0)
6	9.26	9.35	(4 0 0)	11.6	11.6	(3 3 0)
7	8.51	8.58	(3 2 0)	9.24	9.16	(6 3 0)
8	ca.4.6	–	#	ca.4.6	–	#

(*Continued*)

Table 13. *(Continued)*

Entry No.	◎ (91)Hatsusaka3			◎ (92)Hatsusaka4			(93)Hatsusaka6			◎ (94)Hatsusaka7			○ (95)Hatsusaka9			(96)Hatsusaka10		
Reference	p84, $[(C_9O)_2PhO]_8PcCu$: 8a			p84, $[(C_9O)_2PhO]_8PcCu$: 8a			p84, $[(C_{12}O)_2PhO]_8PcCu$: 8d			p84, $[(C_{12}O)_2PhO]_8PcCu$: 8d			p84, $[(C_{12}O)_2PhO]_8PcCu$: 8f			p84, $[(C_{12}O)_2PhO]_8PcCu$: 8f		
Mw	Mw=3588.72			Mw=3588.72			Mw=4262.01			Mw=4262.01			Mw=4710.87			Mw=4710.87		
Temperature	140°C			156°C			120°C			141°C			98°C			122°C		
Mesophase	$Col_{ro}(P2_1/a)$			$Col_{rd}(P2m)$			$Col_{ro}(P2_1/a)$			$Col_{rd}(P2_1/a)$			Col_{hd}			$Col_{rd}(P2_1/a)$		
Lattice constants	a = 66.2, b = 32.4, h_2 = 10.0, Z = 4.0 for ρ = 1.1; h_1 = 5.0, Z = 2.0 for ρ = 1.1			a = 53.2, b = 30.2, Z = 1.0 for h = 3.7[c], ρ = 1.0[c]			a = 73.6, b = 36.7, h_2 = 9.96, Z = 4.0 for ρ = 1.05; h_1 = 4.98, Z = 2.0 for ρ = 1.05			a = 64.7, b = 59.7			a = 43.2			a = 77.8, b = 38.8		
Peak No.	$d_{obs.}$	$d_{calcd.}$	*(hkl)*	$d_{obs.}$	$d_{calcd.}$	*(hkl)*	$d_{obs.}$	$d_{calcd.}$	*(hkl)*	$d_{obs.}$	$d_{calcd.}$	*(hkl)*	$d_{obs.}$	$d_{calcd.}$	*(hkl)*	$d_{obs.}$	$d_{calcd.}$	*(hkl)*
1	33.1	33.1	(2 0 0)	30.2	30.2	(0 1 0)	36.8	36.8	(2 0 0)	32.3	32.3	(2 0 0)	37.4	37.4	(1 0 0)	38.9	38.9	(2 0 0)
2	29.1	29.1	(1 1 0)	26.6	26.6	(2 0 0)	33.7	32.9	(1 1 0)	29.9	29.9	(0 2 0)	21.6	21.6	(1 1 0)	36.0	34.7	(1 1 0)
3	13.2	13.0	(3 2 0)	20.0	20.0	(2 1 0)	27.5	26	(2 1 0)	21.9	21.9	(2 2 0)	18.7	18.7	(2 0 0)	21.4	21.6	(3 1 0)
4	11.2	11.0	(6 0 0)	11.5	11.5	(3 2 0)	20.0	20.4	(3 1 0)	12.3	12.3	(3 4 0)	14.2	14.1	(2 1 0)	18.1	17.4	(4 1 0)
5	10.2	10.2+10.0	(5 2 0)+h_2	8.52	8.51	(6 1 0)	14.6	14.6	(3 2 0)	9.05	9.04	(3 6 0)	10.7	10.8	(2 2 0)	15.3	15.5	(3 2 0)
6	9.64	9.69	(3 3 0)	ca.4.6	–	#	13.6	13.7	(5 1 0)	ca.4.6	–	#	9.26	9.35	(4 0 0)	11.6	11.6	(3 3 0)
7	8.15	8.16	(7 2 0)				12.9	13.0	(4 2 0)				8.51	8.58	(3 2 0)	9.24	9.16	(6 3 0)
8	7.32	7.27	(4 4 0)	c: assumed value			11.0	11.0	(3 3 0)				ca.4.6	–	#	ca.4.6	–	#
9	6.40	6.44	(1 5 0)				9.96	9.96	(0 0 1): h_2									
10	5.92	5.91	(11 1 0)				8.86	8.91	(2 4 0)									
11	ca.5.0	5.0	(0 0 2): h_1				8.58	8.6	(3 4 0)									
12	ca.4.5	–	#				7.87	7.87	(5 4 0)									
13							7.36	7.35	(6 4 0)									
14							4.98	4.98	(0 0 2): h_1									
15							ca.4.5	–	#									

(Continued)

Table 13. (*Continued*)

Entry No.	(103)Ariyoshi6			(104)Ariyoshi8			(105)Ariyoshi9			○ (106)Ariyoshi10			(107)Ariyoshi11			○ (108)Ariyoshi12		
Reference	p146, $C_{14}(OC_{12}OH)PcCu$: 5c			Master Thesis, $[C_{12}(Acryloyl)]PcCu$: 6a			Master Thesis, $[C_{12}(Acryloyl)]PcCu$: 6a			Master Thesis, $[C_{12}(Acryloyl)]PcCu$: 6a			Master Thesis, $[C_{12}(Acryloyl)]PcCu$: 6a			Master Thesis, $[C_{14}(Acryloyl)]PcCu$: 6b		
Mw	Mw=3907.44			Mw=3624.85			Mw=3624.85			Mw=3624.85			Mw=3624.85			Mw=3961.49		
Temperature	137°C			120°C			145°C			152°C			175°C			120°C		
Mesophase	Col_{hd}			Col_{hd}			Col_{hd}			Col_{hd}			$Col_{tet\text{-}o}$			Col_{hd}		
Lattice constants	a = 41.5			a = 40.1			a = 39.5			a = 37.8			a = 43.0, h = ca. 3.6; Z = 1.8 for ρ = 0.90			a = 43.2		
Peak No.	$d_{obs.}$	$d_{calcd.}$	(*hkl*)	$d_{obs.}$	$d_{calcd.}$	(*hkl*)	$d_{obs.}$	$d_{calcd.}$	(*hkl*)	$d_{obs.}$	$d_{calcd.}$	(*hkl*)	$d_{obs.}$	$d_{calcd.}$	(*hkl*)	$d_{obs.}$	$d_{calcd.}$	(*hkl*)
1	35.9	35.9	(1 0 0)	34.[illegible]	34.8	(1 0 0)	34.2	34.2	(1 0 0)	32.7	32.7	(1 0 0)	30.4	30.4	(1 1 0)	37.4	37.4	(1 0 0)
2	21.0	20.7	(1 1 0)	20.2	20.1	(1 1 0)	19.8	19.8	(1 1 0)	19.2	18.9	(1 1 0)	21.8	21.5	(2 0 0)	21.6	21.6	(1 1 0)
3	18.2	18.0	(2 0 0)	17.5	17.4	(2 0 0)	17.4	17.1	(2 0 0)	16.7	16.3	(2 0 0)	ca.4.7	–	#	18.9	18.7	(2 0 0)
4	13.8	13.6	(2 1 0)	13.3	13.1	(2 1 0)	13.3	12.9	(2 1 0)	12.6	12.4	(2 1 0)	ca.3.6	–	(0 0 1)	14.2	14.1	(2 1 0)
5	12.3	12.0	(3 0 0)	11.3	11.6	(2 2 0)	9.63	9.49	(3 1 0)	9.07	9.07	(3 1 0)				10.4	10.4	(3 1 0)
6	10.2	10.4	(2 2 0)	9.85	10.0	(3 0 0)	8.72	8.55	(4 0 0)	ca.4.6	–	#				8.68	8.58	(3 2 0)
7	9.24	8.98	(4 0 0)	8.07	7.97	(3 2 0)	8.02	7.85	(3 2 0)							ca.4.5	–	#
8	8.37	8.24	(3 2 0)	ca.4.5	–	#	ca.4.6	–	#									
9	ca.4.6	–	#															
10																		
11																		
12																		
13																		
14																		
15																		

(*Continued*)

Table 13. *(Continued)*

Entry No.	(109)Ariyoshi13			○ (110)Ariyoshi14			(Additional example 1)			◎(Additional example 2)			◎(Additional example 3)			(114) Pelzl Add 4		
Reference	Master Thesis, [C_{14}(Acryloyl)]PcCu: 6b			Master Thesis, [C_{14}(Acryloyl)]PcCu: 6b			p55, $(C_{12})_8$PcCu: 1b Komatsu			p 30, (2-EtC_6O)$_8$PcCu Watanabe			TLT-6 Yelamaggad			DS12 Pelzl		
Mw	Mw=3961.49			Mw=3961.49			Mw=3893.33			Mw=1481.71			Mw=1697.25			Mw=1530.14		
Temperature	133°C			160°C			RT						160°C			130°C		
Mesophase	$Col_{h.d}$			$Col_{tet.o}$			$Col_{ob.d}$			$Col_{ob.d}$			$Col_{ob.o}$			$Col_{ob.d}$		
Lattice constants	a = 42.5			a = 46.1, h = ca. 3.5; Z = 1.0 for ρ = 0.90			a = 25.7, *b* = 22.5, γ = 96.0°			a = 23.1, *b* = 18.9, γ = 103°			a = 33.94, *b* = 31.80, γ = 104.5°, h_l = 3.38; Z = 1. 00 for ρ = 0.80			a = 45.93, *b* = 42.04, γ = 118.9°, *h* = 3.5; Z = 1. 98 for ρ = 0.85		
Peak No.	$d_{obs.}$	$d_{calcd.}$	*(hkl)*	$d_{obs.}$	$d_{calcd.}$	*(hkl)*	dobs.	$d_{calcd.}$	*(hkl)*	$d_{obs.}$	$d_{calcd.}$	*(hkl)*	$d_{obs.}$	$d_{calcd.}$	*(hkl)*	$d_{obs.}$	$d_{calcd.}$	*(hkl)*
1	36.8	36.8	(1 0 0)	32.6	32.6	(1 1 0)	25.7	25.5	(1 0 0)	22.49	22.51	(1 0 0)	32.91	32.86	(1 0 0)	40.2	40.2	(1 0 0)
2	21.4	21.2	(1 1 0)	23.1	23.1	(2 0 0)	18.4	18.3	(1 $\underline{1}$ 0)	18.42	18.42	(0 1 0)	30.77	30.78	(0 1 0)	36.8	36.8	(0 1 0)
3	18.5	18.4	(2 0 0)	ca.4.6	–	#	16.1	16.5	(1 1 0)	16.13	16.14	(1 $\underline{1}$ 0)	26.13	25.94	(1 $\underline{1}$ 0)	22.3	22.3	(1 1 0)
4	14.1	13.9	(2 1 0)	ca.3.5	–	h: (0 0 1)	12.6	12.8	(2 0 0)	13.06	12.90	(1 1 0)	16.41	16.43	(2 0 0)	20.1	20.1	(2 0 0)
5	12.3	12.3	(3 0 0)				11.2	11.8	(0 2 0)	8.74	8.77	(2 1 0)	12.36	12.76	(1 2 0)			
6	10.7	10.6	(2 2 0)				10.8	10.8	(2 1 0)	7.96	7.92	(1 2 0)	10.92	10.95	(3 0 0)			
7	10.2	10.2	(3 1 0)				8.67	8.51	(3 0 0)				4.59	–	#			
8	9.32	9.20	(4 0 0)				7.61	7.75	(3 1 0)				3.38	–	h: (0 0 1)			
9	ca.4.6	–	#				6.11	6.39	(2 3 0)									
10							ca. 4.2	–	#									
11																		
12																		
13																		
14																		
15																		

G. Pelzl et al., Liq. Cryst., 1995, 19, 541.

(Continued)

Table 13. (*Continued*)

Entry No.	(115) K Yamamoto Add 5			(116) K Yazawa Add 6		
Reference	Blockcopolymer MS17			Spider silk_Female spider Trichonephila clavata		
Mw				Mw = 230000		
Temperature	RT			RT		
Mesophase	Col_h			$Col_{ro}(P2_1/a)$		
Lattice constants	a = 31.97			a = 15.34, b = 10.78; c = 3600 for Z = 2.00 and ρ = 1.29		
Peak No.	$d_{obs.}$	$d_{calcd.}$	*(hkl)*	$d_{obs.}$	$d_{calcd.}$	*(hkl)*
1	27.69	27.69	(1 0 0)	8.82	8.82	(1 0 0)
2	15.96	15.99	(1 1 0)	5.39	5.39	(0 2 0)
3	13.42	13.85	(2 0 0)	4.39	4.41	(2 2 0)
4	9.925	10.47	(2 1 0)	4.10	-	*
5				3.67	3.71	(3 2 0)
6				3.48	3.50	(1 3 0)
7				2.89	2.94	(3 3 0)
8				2.67	2.67	(5 2 0)
9				2.21	2.21	(4 4 0)
10				2.06	2.08	(2 5 0)
11				1.86	1.85	(6 4 0)

K. Yamamoto et al., ACS Omega 2017, 2, 8580. The observed spacings were read by using WebPlotDigitizer.

K. Yazawa and K. Ohta, Liq. Cryst., 2023, in press.

* This peak was tentatively assigned as the repetitive distance of amino acid residues in the protein backbone.

References cited in Table 13.

p30. K. Ohta, T. Watanabe, H. Hasebe, Y. Morizumi, T. Fujimoto, I. Yamamoto, D. Lelievre and J. Simon *Mol. Cryst. Liq. Cryst.*, **196**, 13–26 (1991).

p52. M. Ikejima, M. Moriya, H. Hasebe, T. Fujimoto, I. Yamamoto and K. Ohta, in "*Chemistry of Functional Dyes*" ed. by Z. Yoshida, Y. Shirota, Mita Press, Tokyo, Vol.**2**, pp. 801–805 (1993).

p55. T. Komatsu, K. Ohta, T. Fujimoto and I. Yamamoto, *J. Mater. Chem.*, **4**, 533–536 (1994).

p56. K. Ohta, M. Moriya, M. Ikejima, H. Hasebe, T. Fujimoto and I. Yamamoto, *Bull. Chem. Soc., Jpn.*, **66**, 3553–3558 (1993).

p60. K. Ohta, S. Azumane, T. Watanabe, S. Tsukada and I. Yamamoto, *Appl. Organomet.*, **10**, 623–635 (1996).

p67. K. Ohta, R. Higashi, M. Ikejima, I. Yamamoto and N. Kobayashi, *J. Mater. Chem.*, **8**, 1979–1991 (1998).

p68. K. Ohta, N. Yamaguchi and I. Yamamoto, *J. Mater Chem.*, **8**, 2637–2650 (1998).

p72. H. Miwa, N. Kobayashi, K. Ban and K. Ohta, *Bull. Chem. Soc., Jpn.*, **72**, 2719–2728 (1999).

p74. K. Ohta, S. Azumane, N. Kobayashi and I. Yamamoto, *J. Mater Chem.*, **9**, 2313–2320 (1999).

p75. K. Ohta, Y. Inagaki-Oka, H. Hasebe and I. Yamamoto, *Polyhedron*, **19**, 267–274(2000).

p78. K. Ban, K. Nishizawa, K. Ohta and H. Shirai, *J. Mater Chem.*, **10**, 1083–1090 (2000).

p83. K. Ban, K. Nishizawa, K. Ohta, A. M. van de Craats, J. M. Warman, I. Yamamoto and H. Shirai, *J. Mater. Chem.*, **11**, 321–331 (2001).

p84. K. Hatsusaka, K. Ohta, I. Yamamoto and H. Shirai, *J. Mater. Chem.*, **11**, 423–433 (2001).

p92. T. Nakai, K. Ban, K. Ohta and M. Kimura, *J. Mater. Chem.*, **12**, 844–850 (2002).

p146. M. Ariyoshi, M. Sugibayashi-Kajita, A. Suzuki-Ichihara, T. Kato, T. Kamei, E. Itoh and K. Ohta, *J. Porphyrins Phthalocyanines*, **16**, 1114–1123 (2012).

p182. K. Itoh, N. Kobayashi and K. Ohta, *J. Porphyrins Phthalocyanines*, **25**, 188–201 (2021).

Kaori Nihsizawa, Bachelor Thesis, Shinshu University, Ueda, Japan, 1997.

Motoko Aoki, Bachelor Thesis, Shinshu University, Ueda, Japan, 1995.

Koh-ichi Hatada, Master Thesis, Shinshu University, Ueda, Japan, 1987.

Kazue Ban, Master Thesis, Shinshu University, Ueda, Japan, 1999.

Masaaki Ariyoshi, Master Thesis, Shinshu University, Ueda, Japan, 2001.

J. Billard, J. C. Dubois, C. Vaucher and A. M. Levelut, *Mol. Cryst. Liq. Cryst.*, 66, 115–122 (1981).

A. S. Achalkumar, U. S. Hiremath, D. S. S. Rao, S. K. Prasad and C. V. Yelamaggad, *J. Org. Chem.*, **78**, 527–544 (2013). [For Sample (113)]

W. Weissflog, M. Rogunova, I. Letko, S. Diele and G. Pelzl, *Liq. Cryst.*, **19**, 541–544 (1995). [For Sample (114)]

J. Fukuhara, A. Yasui, K. Yamamoto and S. Sakurai, ACS Omega, **2**, 8580–8590 (2017). The observed spacings were read by using WebPlotDigitizer. [For Sample (115)]

K. Yazawa and K. Ohta, *Liq. Cryst.*, (2023) in press. [For Sample (116)]

Additionally, the mesophases of M1∼M11 listed in the textbook were also analyzed by using Bunseki-kun Ver.3. The X-ray data are also listed in the following.

Table 14. X-ray data and the identifications of mesophases for Sample M1∼M11. The liquid crystalline phases were analyzed by "Reciprocal Lattice Method" developed by us.

Entry No.	M1			M2			M3			M4			M5			M6		
Reference	1, Ni(12,I⁻)			10, $[(C_{14}O)_2PhO]_8PcCu$			7, $[(C_{18}OPh)_8Pc]_2Lu$			5, $(C_{12}OPzCu$			36, C_{10}-Cu			9, RHO		
Mw	Mw=1330.61			Mw=4710.87			Mw=6713.33			Mw=1930.58			Mw=3685.15			Mw=1061.416		
Temperature	170°C			98°C			RT			150°C			100°C			T in Dc		
Mesophase	Col_{ho}			Col_{hd}			$Col_{tet.d}$			$Col_{rd}(C2/m)$			$Col_{rd}(P2_1/a)$			$Col_{rd}(P2/a)$		
Lattice constants	a=28.5, h=3.36; Z=1.07 for ρ=1.00			a=29.95, h=3.82; Z=1.0 for ρ=1.20			a=34.8			a=57.8, b=29.5, h=3.4; Z=1.99 for ρ=1.10			a=56.6, b=38.8, h=5.0; Z=2.0 for r=1.1			a=35.7, b=34.9; Z=4.0 for ρ=1.0		
Peak No.	$d_{obs.}$	$d_{calcd.}$	*(hkl)*	$d_{obs.}$	$d_{calcd.}$	*(hkl)*	$d_{obs.}$	$d_{calcd.}$	*(hkl)*	$d_{obs.}$	$d_{calcd.}$	*(hkl)*	$d_{obs.}$	$d_{calcd.}$	*(hkl)*	$d_{obs.}$	$d_{calcd.}$	*(hkl)*
1	24.7	24.7	(1 0 0)	37.4	29.9	(1 0 0)	38.8	30.1	(1 0 0)	28.9	28.9	(2 0 0)	32.0	32.0	(1 1 0)	25.0	25.0	(1 1 0)
2	14.2	14.3	(1 1 0)	21.6	21.2	(1 1 0)	28.4	17.4	(1 1 0)	26.3	26.3	(1 1 0)	28.3	28.3	(2 0 0)	17.5	17.5	(0 2 0)
3	12.3	12.4	(2 0 0)	18.7	9.47	(1 3 0)	20	15.1	(2 0 0)	16.4	16.1	(3 1 0)	18.6	18.4	(1 2 0)	16.0	15.9	(2 1 0)
4	9.24	9.34	(2 1 0)	14.2	8.31	(2 3 0)	14.4	11.4	(2 1 0)	14.7	14.8	(0 2 0)	12.6	12.9	(1 3 0)	9.76	9.75	(2 3 0)
5	*ca.* 4.8	–	#	10.7	7.26	(4 1 0)	13.1	10.0	(3 0 0)	13.2	13.2	(2 2 0)	11.2	11.4	(4 2 0)	8.98	8.92	(4 0 0)
6	3.36	3.36	*h*: (0 0 1)	9.26	6.70	(2 4 0)	*ca.* 4.2	8.70	(2 2 0)	*ca.* 4.7	–	#	*ca.* 4.6	–	#	*ca.* 5.6	broad	h_1: (0 0 1)
7				8.51	4.99	(6 0 0)				*ca.* 3.4	*ca.* 3.4	*h*: (0 0 1)				*ca.* 4.5	–	#
8				*ca.* 4.6	–	#												
9																		
10																		
11																		
12																		
13																		
14																		
15																		

Table 14. *(Continued)*

Entry No.	M7			M8 = (111)Komatsu Add. 1			M9			M10			M11		
Reference	7, $(2\text{-Et-}C_6)_8PcH_2$			p55, $(C_{12}Pc)_2Lu$			X(p45), $(C_{12}Salen)_2Ni$			X(p54), 10(n=80)			p76, $(C_6)_2DABCO\text{-}Br_2$		
Mw	Mw=1414.288			Mw=3893.33			Mw=633.57			Mw=380.53			Mw=722.86		
Temperature	55°C			RT			200°C			122°C			120°C		
Mesophase	$Col_{rd}(P2m)$			$Col_{ob.d}$			S_A			$S_E(P2_1/a)$			S_T		
Lattice constants	a = 70.8, b = 32.6			a = 25.7, b = 22.5, g = 96.0°			c = 33.3			a = 8.12, b = 5.48, c = 25.7 Z = 2.0 for ρ = 1.1			a = 6.21, b = 6.21, c = 32.0 Z = 1.0 for ρ = 1.0		
Peak No.	$d_{obs.}$	$d_{calcd.}$	*(hkl)*	$d_{obs.}$	$d_{calcd.}$	*(hkl)*	$d_{obs.}$	$d_{calcd.}$	*(hkl)*	$d_{obs.}$	$d_{calcd.}$	*(hkl)*	$d_{obs.}$	$d_{calcd.}$	*(hkl)*
1	20.4	20.4	(1 0 0)	25.7	25.7	(1 0 0)	32.8	32.8	(0 0 1)	25.2	25.7	(0 0 1)	32.9	32.9	(0 0 1)
2	18.6	18.6	(0 1 0)	13.4	17.8	(1 1 0)	16.7	16.6	(0 0 2)	13.1	12.9	(0 0 2)	16.1	16.0	(0 0 2)
3	13.6	13.7	(1 1 0)	16.1	16.1	(1 1 0)	11.2	11.1	(0 0 3)	4.54	4.54	(1 1 0)	10.7	10.7	(0 0 3)
4	10.2	10.2	(2 0 0)	12.6	12.9	(2 0 0)	*ca.* 4.9	–	#	4.06	4.06	(2 0 0)	7.97	8.00	(0 0 4)
5	8.56	8.45	(1 2 0)	11.2	11.2	(0 2 0)				3.26	3.26	(2 1 0)	6.35	6.40	(0 0 5)
6	6.78	6.80	(3 0 0)	10.8	10.7	(2 1 0)							5.28	5.34	(0 0 6)
7	6.13	6.19	(0 3 0)	8.67	8.58	(3 0 0)							4.51	4.57	(0 0 7)
8	5.89	5.92	(1 3 0)	7.61	7.73	(3 1 0)							4.40	4.39	(1 1 0)
9	5.22	5.29	(2 3 0)	6.11	6.17	(2 3 0)							3.10	3.10	(2 0 0)
10	5.07	5.10	(4 0 0)	*ca.* 4.2	–	#							2.77	2.78	(1 2 0)
11	4.87	4.92	(4 1 0)												
12	4.66	4.64	(0 4 0)												
13	4.50	4.52	(1 4 0)												
14	4.42	4.47	(4 2 0)												
15	4.25	4.22	(2 4 0)												
16	4.08	4.08	(5 0 0)												
17	3.85	3.84	(3 4 0)												
18	3.70	3.71	(0 5 0)												
19	3.42	3.43	(4 4 0)												
20	3.11	3.09	(0 6 0)												
21	3.01	3.00	(4 5 0)												
22	*ca.* 4.4	–	#												

#: halo of molten alkyl chains

Table 15. Classification of the liquid crystal structures and dimensionalities for the samples of (1)~(114) in Table 13 and M1~M11 in Table 14.

Mesophase	Z value	Dimension	Entry No.
Col_{ho}	1	2D⊕1D	○**M1**,1,2,3,8,30,31,32,33,34,35,36,37,51,52,53,54,58,59,60, 61,63,64, 65,66,68,70,71,72,73,74,76,78,79,80,82,84,85,86,87
Col_{hd}	1#	2D	M2,7,21,24,88,89,95,97,98,100,102,103,104,105,106,108,109
$Col_{tet.o}$	1	2D⊕1D	15,18,○**20**,○**99**,101,107,110
$Col_{tet.d}$	1#	2D	M3, 12,16,62
Col_{ro}(P2m)	1	2D⊕1D	38
Col_{rd}(P2m)	1#	2D	○**M7**, 92
Col_{ro}($P2_1/a$)	2	2D⊕1D	29,90,91,93
Col_{rd}($P2_1/a$)	2#	2D	○**M5**, 5,9,10,11,13,14,17,42,43,94,96
Col_{ro}(C2/m)	2	2D⊕1D	Non
Col_{rd}(C2/m)	2#	2D	○**M4**, 4,6,19,22,23,25,26,28
Col_{ro}(P2/a)	4	2D⊕1D	○**M6**, 49,55,57
Col_{rd}(P2/a)	4#	2D	Non
Pseudo-Col_{ho}	1 and 2	2D⊕1D	67,69,75, ○**77**,81,83
Pseudo-Col_{hd}	1# and 2#	2D	Non
D_{L1}	X	1D	44,45,46,47
D_{L2} (D_{LC}=Col_{L})	X	1D⊕1D	27,39, 40, 41,48,50,56
$Col_{ob.o}$	1#	2D⊕1D	○**113**
$Col_{ob.d}$	1	2D	**M8**(=111), 112, 114
S_A	X	1D	○**M9**
S_E($P2_1/a$)	2	2D⊕1D	○**M10**
S_T	1	2D⊕1D	○**M11**

X = The Z value is impossible to be calculated.

= If the stacking distance in these disordered mesophases is assumed, the Z value can be calculated.

○: The sample selected for the analysis procedure explanation using the Bunseki-kun program.

Finally, the author introduces a very useful non-3D phase structure analysis computer program for soft matter, named as "**Bunseki-kun Ver. 3**".

If you use only a calculator, graph paper, and a ruler, it would probably take more than one year to analyze all of the phase structures in the 116 soft matter materials listed in Table 13. In order to shorten the analysis time, our research laboratory group has developed a computer program and polished up to Bunseki-kun Ver.3. If you will use this program, you will be able to easily analyze the soft matter phase structures within 1/100th to 1/1000th of the time. Accordingly, you are strongly recommended to utilize this computer program to greatly shorten the analysis time. For those who purchase this book, you will be able to freely download this program by following the instructions given in the next page.

For your easy comprehension, this book provides the representative selected examples and their detailed analysis manners as Part II: "How to use the X-ray non-3D phase structure analysis program, Bunseki-kun Ver.3". By using both the computer program and the manual, you will be able to acquire the X-ray soft matter structure analysis technique in short time.

Supplementary Material (14 MB)

If you have purchased the print copy of this book or the ebook online via other sales channels, please follow the instructions below to download the files:

Register an account/login at https://www.worldscientific.com

Go to: https://www.worldscientific.com/r/13310-supp

Download the files from: https://www.worldscientific.com/worldscibooks/10.1142/13310#t=suppl. For subsequent access, simply log in with the same login details in order to access.

For enquiries, please email: sales@wspc.com.sg.

Part II

How to Use the X-ray Non-3D Phase Structure Analysis Program, "Bunseki-kun Ver.3"

1 Outline of "Bunseki-kun" and Common Procedure for Data Saving and Correction

1.1 Introduction

The presently published book is the solution book for the end-of-chapter problems of the previously published "Physics and Chemistry of Molecular Assemblies". This solution book contains the correct solutions for the end-of-chapter problems of Chapters 1 to 3. In particular, the 110 problems for X-ray liquid crystal structure analysis in Problem 9 of Chapter 3 is so huge. These X-ray liquid crystal structural analyses can be basically done with "**Reciprocal Lattice Method**" described in the previous book. However, it will probably take one year to solve all the problems, if you use only a calculator, graph paper, a ruler, and a compass.

Therefore, we, Shinshu University Ohta Laboratory, developed a computer program "**Bunseki-kun** Ver.1" for easy analysis in 1993, and the program has been improved several times, so that we can analyse the liquid crystal structures and soft matter in an extremely short time. With this program, the analysis can be carried out in 1/100 to 1/1000 of the time, in comparison with the method using graph paper. Such a computer program for X-ray liquid crystal structure analysis and soft matter has never been published in the whole world to date to our best knowledge.

"**Bunseki-kun**" can be applied not only to the liquid crystalline phases, but also to all kinds of soft matter phases having the dimensionalities of 1D, 2D, 1D⊕1D, 2D⊕1D, and 1D⊕1D⊕1D, except for 3D. Accordingly, it can be applied to thermotropic liquid crystals such as rod-like liquid crystals, discotic liquid crystals, and flying-seed liquid crystals, lyotropic liquid crystals, low-molecular-weight liquid crystals, high-molecular-weight liquid crystals, biopolymers, block copolymers, and flexible fibres such as silk and spider silk resulting from liquid crystal spinning.

This program was originally written in the N88 Basic language in 1993, but the NEC PC98 personal computers that can run N88 Basic are no longer available on the market. In 2021–2022, we therefore translated this program to Visual Basic, which runs on the current Windows PC. During the translation, we received kind support from the Yasutake Laboratory of Saitama University.

For those who have purchased this book, this program "**Bunseki-kun** Ver.3" is available for download free of charge from the publisher's website, together with this manual entitled as "How to use the liquid crystal structure analysis program, **Bunseki-kun** Ver3." Instructions to download this program is available in the Supplementary Material.

The temperature-dependent X-ray diffraction data for the 110 exercises in Problem 9 of Chapter 3, are mainly for discotic liquid crystal phases newly synthesized in Ohta laboratory, Shinshu University, but these analysis methods can be applied to other soft matter in general. Hence, the author expects that this computer program will be widely used for the structural analysis of all kinds of soft matter. The author also hopes that this program will be used not only for the exercises but also for your actual research.

In this solution book, we have added four new exercises for the $\mathrm{Col_{ob}}$ phase, which were not mentioned in the previous textbook. The reason why the author didn't include these exercises is that the $\mathrm{Col_{ob}}$ liquid crystal phase cannot be solved only with a calculator, graph paper, a ruler, and a compass; they cannot be solved without programming a computer application.

When the author discloses this program of **Bunseki-kun** Ver.3, he includes the additional six problems of the $\mathrm{Col_{ob}}$ phase, the $\mathrm{Col_h}$ phase in a **block copolymer** and the $\mathrm{Col_{ro}}$ phase in a **spider silk**, and wishes to explain their representative X-ray liquid crystal structure analyses selecting from all the (110+6) problems. The 116 problems in Problem 9 include liquid crystalline phases with various dimensions. Table 1 summarizes the appeared types and dimensionality of the liquid crystal phases.

Table 1. Summary of the (110 + 6) mesophases in Table 13 and M1 ~ M11 in Table 3.12 in the textbook.

Mesophase	Z value	Dimension	Entry No.
$\mathrm{Col_{ho}}$	1	2D⊕1D	○**M1**,1,2,3,8,30,31,32,33,34,35,36,37,51,52,53,54,58,59,60, 61,63,64, 65,66,68,70,71,72,73,74,76,78,79,80,82,84,85,86,87
$\mathrm{Col_{hd}}$	1#	2D	M2,7,21,24,88,89,95,97,98,100,102,103,104,105,106,108,109, 115
$\mathrm{Col_{tet.o}}$	1	2D⊕1D	15,18,○**20**,○**99**,101,107,110
$\mathrm{Col_{tet.d}}$	1#	2D	M3, 12,16,62
$\mathrm{Col_{ro}}$(P2m)	1	2D⊕1D	38
$\mathrm{Col_{rd}}$(P2m)	1#	2D	○**M7**, 92
$\mathrm{Col_{ro}}$(P2$_1$/a)	2	2D⊕1D	29,90,91,93, 116
$\mathrm{Col_{rd}}$(P2$_1$/a)	2#	2D	○**M5**, 5,9,10,11,13,14,17,42,43,94,96
$\mathrm{Col_{ro}}$(C2/m)	2	2D⊕1D	Non
$\mathrm{Col_{rd}}$(C2/m)	2#	2D	○**M4**, 4,6,19,22,23,25,26,28
$\mathrm{Col_{ro}}$(P2/a)	4	2D⊕1D	○**M6**, 49,55,57
$\mathrm{Col_{rd}}$(P2/a)	4#	2D	Non
Pseudo-$\mathrm{Col_{ho}}$	1 and 2	2D⊕1D	67,69,75, ○**77**,81,83
Pseudo-$\mathrm{Col_{hd}}$	1# and 2#	2D	Non
$\mathrm{D_{L1}}$	X	1D	44,45,46,47
$\mathrm{D_{L2}}$ ($\mathrm{D_{LC}}$=$\mathrm{Col_L}$)	X	1D⊕1D	27,39, 40, 41,48,50,56
$\mathrm{Col_{ob.o}}$	1#	2D⊕1D	○**113**
$\mathrm{Col_{ob.d}}$	1	2D	**M8**(=111), 112, 114
$\mathrm{S_A}$	X	1D	○ **M9**
$\mathrm{S_E}$(P2$_1$/a)	2	2D⊕1D	○ **M10**
$\mathrm{S_T}$	1	2D⊕1D	○ **M11**

X = the Z value is impossible to be calculated.
# = If the stacking distance in the disordered mesophases is assumed, the Z value can be calculated.
○ = the sample selected for the analysis procedure explanation using the Bunseki-kun program.

1.2 Outline: Twenty-one types of liquid crystalline phases and the dimensionalities.

As can be seen from this table, 114 mesophases in Problem 9 and the mesophases from M1 to M11 in textbook contain 21 types of liquid crystalline phases. These liquid crystalline phases have 1D, 2D, 1D⊕1D and 2D⊕1D dimensionalities. The present **Bunseki-kun** program using **Reciprocal Lattice Method** can be applied to X-ray structure analysis of any soft matter with these dimensionalities, not limited to columnar liquid crystal phases and smectic liquid crystal phases. Therefore, the author would like all the researchers who study the state of molecular assemblies to utilize the analytical methods presented in this book.

When you analyse the structure of each of these liquid crystal phases, you use the mathematical equations of [2D⊕1D] dimensionality and so on in the following table.

1.3 Outline: Standard peaks for the X-ray liquid crystal structure analysis

Hereupon, when you carry out the structural analysis of a 2D lattice, you consider the standard peak. The standard peaks are basically the strongest reflection peaks appearing on the lowest angle region in the X-ray diffraction pattern. Figure 1 illustrates what the strongest reflections are for each liquid crystal phase.

In this figure, you first look at the Col_h phase. In this Col_h phase, the most densely packed planes are the (10) plane, followed by the (11) plane. Reflections from these surfaces are therefore the strongest. Also, from Table 2, the relationship among the 2D hexagonal lattice spacing d value, lattice constant, and Miller index is (2D) : $\frac{1}{d_{hk0}^2} = \frac{4}{3}\left(\frac{h^2+hk+k^2}{a^2}\right)$. In this equation, there is only one unknown quantity, *i.e.*, the lattice constant a. Therefore, if one plane (10) is selected as a standard peak, the lattice constant a can be calculated from this equation. Using this obtained value a, you can draw the reciprocal lattice plane of the two-dimensional hexagonal lattice. Then, by drawing the Debye-Scherrer rings of all peaks including the standard peak on this reciprocal lattice plane and

Table 2. Relationship among interplanar spacing (d), Miller indices (hkl) and the lattice constants (a, b, c, α, β and γ) for [2D⊕1D] mesophases known to date.

3D	2D⊕1D	Lattices after degradation	Mesophase
Orthorhombic $\frac{1}{d_{hkl}^2}=\frac{h^2}{a^2}+\frac{k^2}{b^2}+\frac{l^2}{c^2}$	⇨ $\begin{cases}\frac{1}{d_{hk0}^2}=\frac{h^2}{a^2}+\frac{k^2}{b^2}\\ \frac{1}{d_{00l}^2}=\frac{l^2}{c^2}\end{cases}$	Rectangular Stacking distance(c = h), Layer thickness (c)	**Col_{ro} (a>b>>c),** **S_E (c>>a>b)**
3D-hexagonal $\frac{1}{d_{hkl}^2}=\frac{4}{3}\left(\frac{h^2+hk+k^2}{a^2}\right)+\frac{l^2}{c^2}$	⇨ $\begin{cases}\frac{1}{d_{hk0}^2}=\frac{4}{3}\left(\frac{h^2+hk+k^2}{a^2}\right)\\ \frac{1}{d_{00l}^2}=\frac{l^2}{c^2}\end{cases}$	2D-hexagonal Stacking distance (c = h), Layer thickness (c)	**Col_{ho} (a>>c),** **S_B (c>>a)** **(S_L)**
3D-tetragonal $\frac{1}{d_{hkl}^2}=\frac{h^2+k^2}{a^2}+\frac{l^2}{c^2}$	⇨ $\begin{cases}\frac{1}{d_{hk0}^2}=\frac{h^2+k^2}{a^2}\\ \frac{1}{d_{00l}^2}=\frac{l^2}{c^2}\end{cases}$	2D-tetragonal Stacking distance (c = h), Layer thickness (c)	**$Col_{tet.o}$ (a>>c),** **S_T (c>>a)**
Monoclinic α = 90°, β = 90°, γ ≠ 90° a>b>>c $\frac{1}{d_{hkl}^2}=\frac{1}{\sin^2\gamma}\left(\frac{h^2}{a^2}+\frac{k^2}{b^2}-\frac{2hk\cos\gamma}{ab}\right)+\frac{l^2}{c^2}$	The a axis tilts to the b axis direction. ⇨ $\begin{cases}\frac{1}{d_{hk0}^2}=\frac{1}{\sin^2\gamma}\left(\frac{h^2}{a^2}+\frac{k^2}{b^2}-\frac{2hk\cos\gamma}{ab}\right)\\ \frac{1}{d_{00l}^2}=\frac{l^2}{c^2}\end{cases}$ 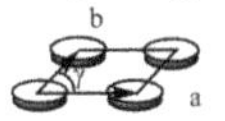 b a (a>b>>c) γ ≠ 90°	Oblique Stacking distance (c = h)	**$Col_{ob.o}$**
Monoclinic α = 90°, β ≠ 90°, γ = 90° c>>a>b $\frac{1}{d_{hkl}^2}=\frac{1}{\sin^2\beta}\left(\frac{h^2}{a^2}+\frac{l^2}{c^2}-\frac{2hl\cos\beta}{ac}\right)+\frac{k^2}{b^2}$	The c axis tilts to the a axis direction. ⇨ $\begin{cases}\frac{1}{d_{hk0}^2}=\frac{h^2}{a^2}+\frac{k^2}{b^2}\\ \frac{1}{d_{00l}^2}=\frac{l^2}{c^2}\end{cases}$ 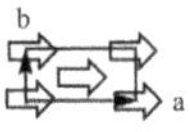 b a (c>>a>b) β ≠ 90°	Rectangular Layer thickness (c)	**S_F** **(S_G)** **(S_H)**
Monoclinic α ≠ 90°, β = 90°, γ = 90° c>>a>b $\frac{1}{d_{hkl}^2}=\frac{1}{\sin^2\alpha}\left(\frac{k^2}{b^2}+\frac{l^2}{c^2}-\frac{2kl\cos\alpha}{bc}\right)+\frac{h^2}{a^2}$	The c axis tilts to the b axis direction. ⇨ $\begin{cases}\frac{1}{d_{hk0}^2}=\frac{h^2}{a^2}+\frac{k^2}{b^2}\\ \frac{1}{d_{00l}^2}=\frac{l^2}{c^2}\end{cases}$ 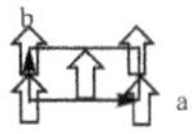 b a (c>>a>b) α ≠ 90°	Rectangular Layer thickness (c)	**S_I** **(S_J)** **(S_K)**

d; spacing; h, k, l: Miller index; a, b, c, α, β, γ: lattice constants.
Definition: α is the angle between b and c axes; β is the angle between a and c axes; γ is the angle between a and b axes.

reading the lattice points at the intersections passed by the rings, the reflection peaks can be indexed. Thus, one reflection peak from the (10) plane is required as a standard peak to analyse the Col_h phase.

Similarly, when you also analyse the Col_{tet} phase, one reflection peak from the (10) plane is required as a standard peak.

When you next look at the $Col_r(C2/m)$ phase in Figure 1, the strongest reflections are from the (10) plane, then the (11) plane. However, the (10) reflection peak does not appear due to the extinction rule. Therefore, the next strongest reflection is from the

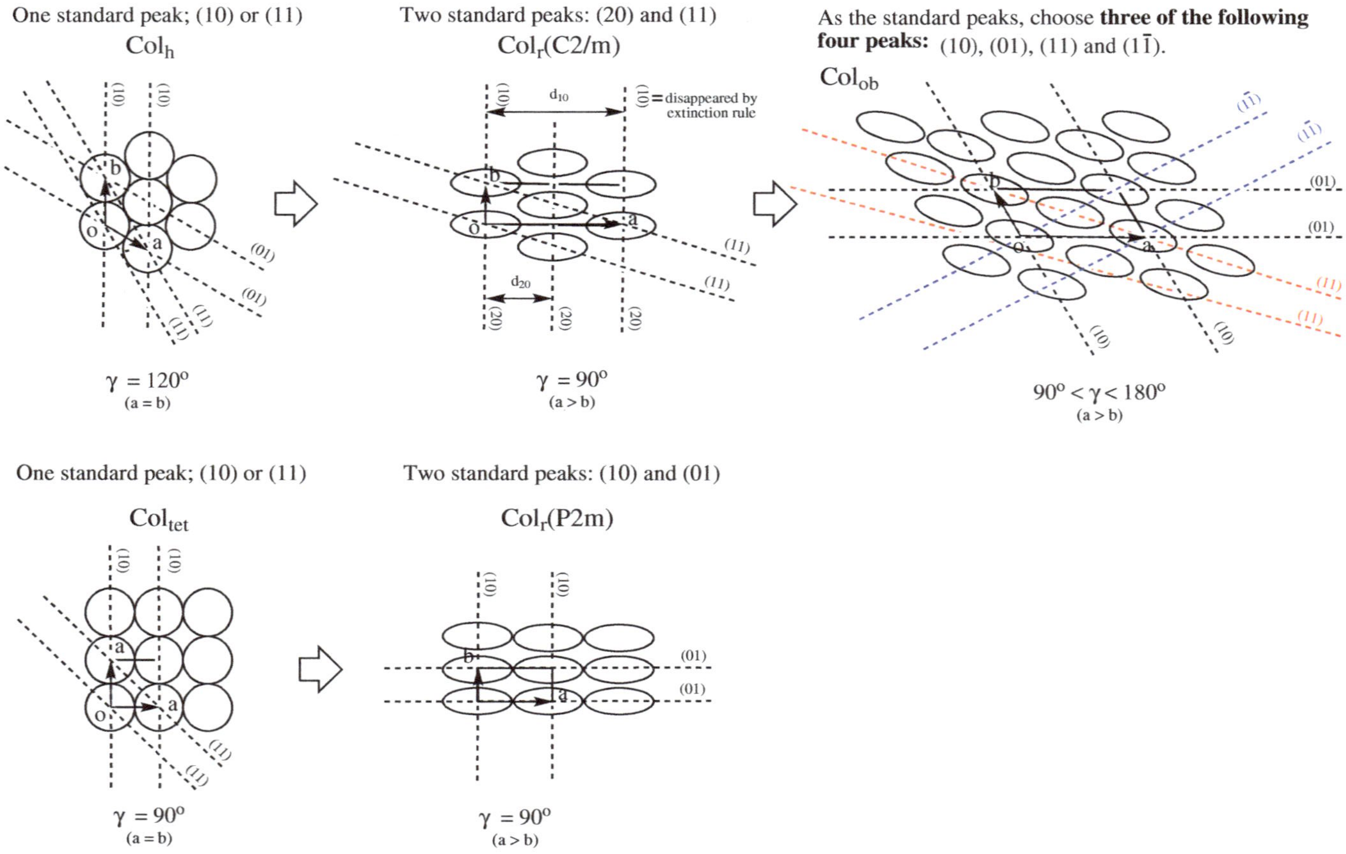

Figure 1. How many standard peaks are needed for the X-ray liquid crystalline phase structural analysis?

(20) plane. It is apparent from Table 2 that the relationship among the spacing d value, the lattice constant, and the Miller index of the 2D rectangular lattice is (2D) : $\frac{1}{d_{hk0}^2} = \frac{h^2}{a^2} + \frac{k^2}{b^2}$. In this equation, the two unknown quantities are the lattice constants a and b. Therefore, when the reflection peaks from the (20) plane and the (11) plane are selected as the standard peaks, the lattice constants a and b can be obtained from this equation. Using these a and b, you can draw a reciprocal lattice plane of a two-dimensional rectangular lattice. Then, by drawing the Debye-Scherrer rings of all peaks including the standard peaks on this reciprocal lattice plane and reading the lattice points at the intersections passed by the rings, the reflection peaks can be indexed. Thus, two reflection peaks from the (20) and (11) planes are required as standard peaks to analyse the Colr(C2/m) phase. Also, in the cases of $Col_r(P2_1/a)$ and $Col_r(P2/a)$ phases, two reflection peaks from the (20) and (11) planes should be also selected as the standard peaks.

Similarly, when you analyse the $Col_r(P2m)$ phase, two reflections are required as standard peaks. However, since there is no extinction rule for this symmetry, the reflection peak form the (10) plane appears as a strong reflection. Therefore, two reflection peaks from the (10) and (01) planes are selected as standard peaks.

When you finally look at the Col_{ob} phase in Figure 1, the strongest reflections are from the (10), (01), (11), and ($1\bar{1}$) planes. It can be seen from Table 2 that the relationship among the spacing d value, the lattice constants, and the Miller index for the 2D oblique lattice (parallelogram) is given by $\frac{1}{d_{hk0}^2} = \frac{1}{sin^2\gamma}\left(\frac{h^2}{a^2} + \frac{k^2}{b^2} - \frac{2hkcos\gamma}{ab}\right)$. In this equation, the unknown quantities are the lattice constants a, b, and γ. Therefore, if three of the four reflection peaks from the planes (10), (01), (11), and ($1\bar{1}$) are selected as the standard peaks, the lattice constants a, b, and γ can be obtained from this equation. Using these a, b, and γ, you can draw the reciprocal lattice plane of a two-dimensional parallelogram (oblique lattice). Then, by drawing the Debye-Scherrer rings of all peaks including the standard peaks on this reciprocal lattice plane and reading the lattice points at the intersections passed by the rings, the reflection peaks can be indexed.

1.4 Our developed program for liquid crystal structure analysis, Bunseki-kun, is consist of "Estimation Method of Dimensionality," "Reciprocal Lattice Method" and "FlexiLattice Method"

Thus, it should be possible in principle from the Reciprocal Lattice method to solve the by such a procedure mentioned above. However, it is extremely difficult for the Col_{ob} phase to determine which reflection peaks are originated from (10), (01), (11) and $(1\bar{1})$ planes. Therefore, in this case, it is easier and more accurate to solve by using another analytical technique of "**FlexiLattice Method**" than by "**Reciprocal Lattice Method.**" The details are given at the end of this guide. Our developed program for liquid crystal structure analysis, Bunseki-kun, is consist of "Estimation Method of Dimensionality," "Reciprocal Lattice Method" and "FlexiLattice Method."

Here, the author wishes to explain the details of the liquid crystal structure analysis procedures only for the representative 11 samples by using the program, Bunseki-kun Ver.3. Because, it is not possible

Table 3. Entry number, mesophase, Z value, dimensionality, data name and sample name of representative examples.

Entry No.	Mesophase	Z value	Dimensionality	Data name	Sample name
1	Common procedure for data saving and correction				
2	Col_{ho}	1	2D⊕1D	○M1	Ni(12, 1)
3	Col_{rd}(C2/m)	$2^{\#}$	2D	○M4	C_{12}PzCu
4	Col_{rd}($P2_1$/a)	$2^{\#}$	2D	○M5	C_{10}-Cu
5	Col_{ro}(P2/a)	4	2D⊕1D	○M6	RHO
6	Col_{rd}(P2m)	$1^{\#}$	2D	○M7	$(2\text{-Et-}C_6)_8PcH_2$
7	$Col_{tet.o}$	$1^{\#}$	2D⊕1D	○20,○99	$(C_{10}OPh)_8$PzCu, $C_{14}(OC_{12}OH)$PcCu
8	S_A	X	1D	○M9	$(C_{12}Salen)_2Ni$
9	S_E($P2_1$/a)	2	2D⊕1D	○M10	10(n =80)
10	S_T	1	2D⊕1D	○M11	$(C_6)_2$DABCO-Br_2
11	$Col_{ob.o}$	$1^{\#}$	2D⊕1D	○113	(113)Yelamaggad Add. 3_TLT-6
12	Pseudo-Col_{ho}	1 and 2	2D⊕1D	○77	(77)Ban27_ $[(C_{10}S)_8Pc]_2$Eu: 1c

○ = the sample selected for the analysis procedure explanation using the Bunseki-kun program.
X = the Z value is impossible to be calculated.
# = If the stacking distance in these disordered mesophases is assumed, the Z value can be calculated.

to provide all the details for the (114 + 11 = 125) samples in the Table 3–12 and End of Chapter Problem 9 in Chapter 3 of the textbook.

In Table 3 are listed the samples selected here for the analysis procedure explanation using the **Bunseki-kun** program in this part. The detailed procedures of the liquid crystal structure analysis are described for Entry numbers 2 to 10 and 12 by using the "**Reciprocal Lattice Method**", and for Entry number 11 by using the "**FlexiLattice Method.**" If you refer to the procedures for these representative examples, you can get all the correct answers for M1-11 and (110 + 4) in Problem 9 in the textbook.

1.5 Common procedure for data saving and correction

1.5.1 *Saving X-ray diffraction data into "Bunseki-kun"*

We number the peaks in the X-ray diffraction pattern, e.g., Figure 2, and read each of the spacing values. All units are used in Å in this document.

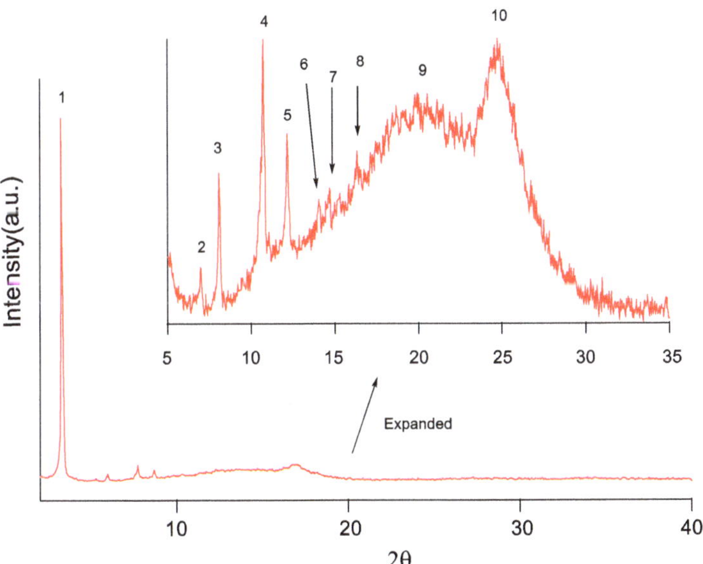

Figure 2. X-ray diffraction pattern of the $[(C_5O)_4DPD]_2Pd$ complex [(8) Oka1] at 130°C.

Step (1) Application selection menu screen in Bunseki-kun

When you open the X-ray soft matter structure analysis program, "BunsekiKunVer.3," the application selection menu screen appears as follows:

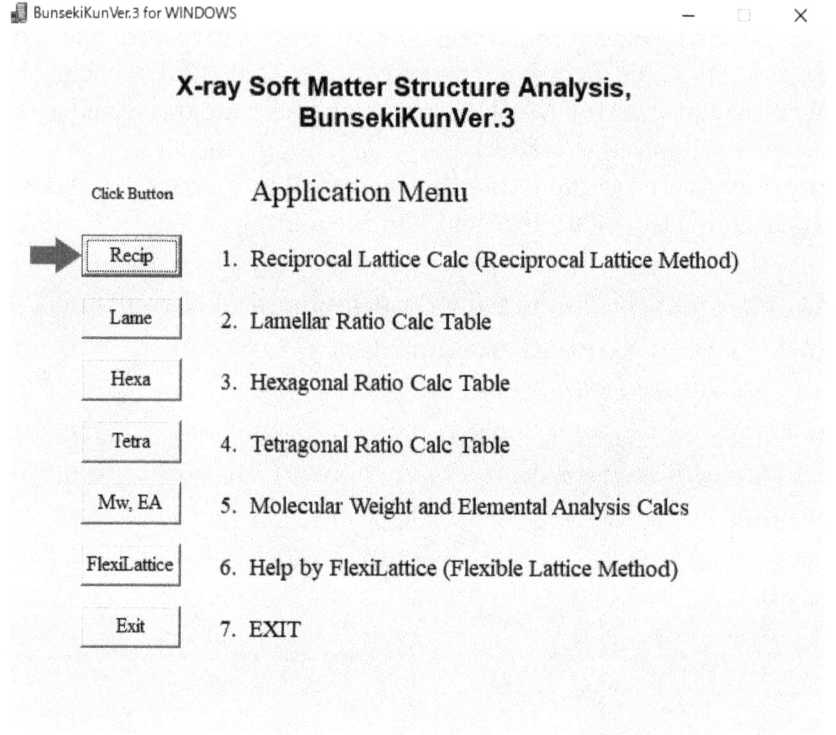

Here, at first, press the [Recip] (reciprocal lattice calculation) button to input the observed values of X-ray diffraction.

Step (2) The initial screen of "Reciprocal Lattice Calculations" appears.

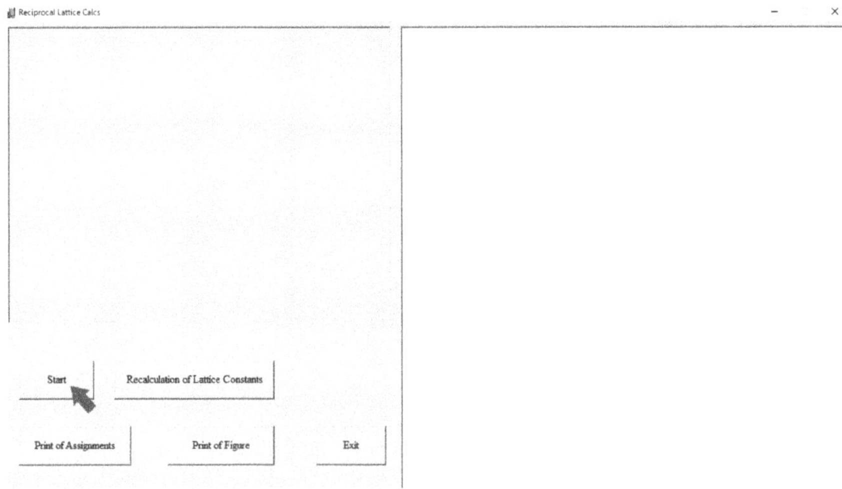

Step (3) When you press the [Start] button, the following selection screen appears.

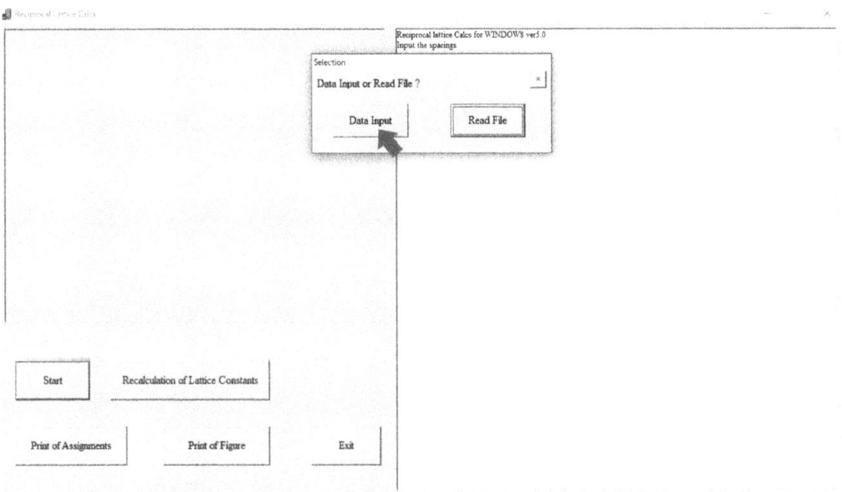

Then press the [Data Input] button. If the hard disk already contains the data, press [Read File]. If there is no data yet, press [Data Input].

Step (4) Enter the number of data

At first, enter the number of observed values.

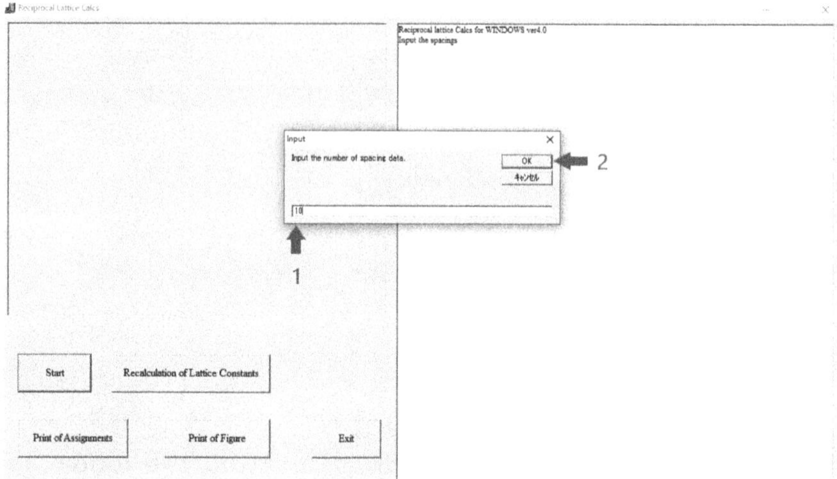

Step (5) Input the observed spacing values

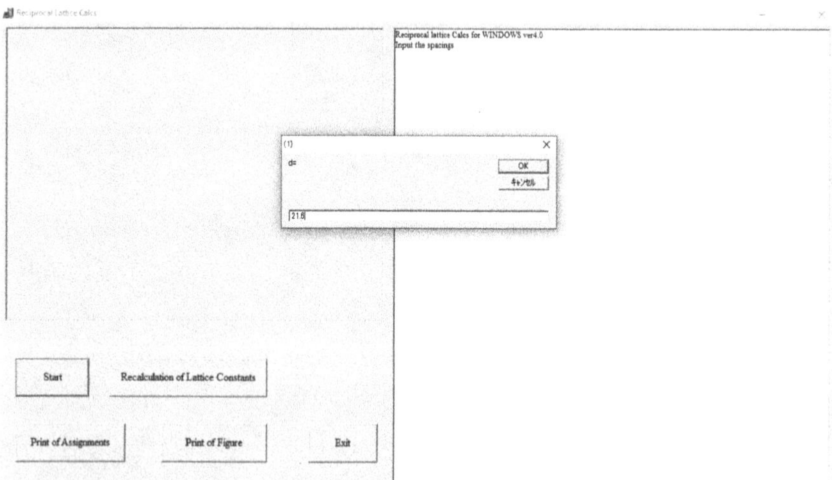

Repeat data input for the number of observed values. When the observed data number is reached, the next screen appears automatically.

1.5.2 *Correction, change, deletion and addition of the data X-ray diffraction data in "Bunseki-kun"*

Step (6) Correction, change, deletion and addition of the data

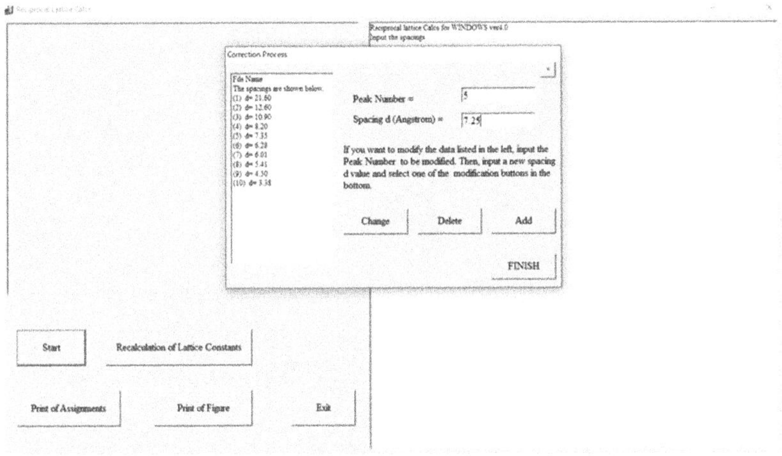

For example, if you mistyped the fifth value, enter the data number, input the correct value, and then press the [Change] button.

Similarly, you can [Delete] and [Add] the data.

Step (7) When the correction is completed or when there is no correction, press the [FINISH] button.

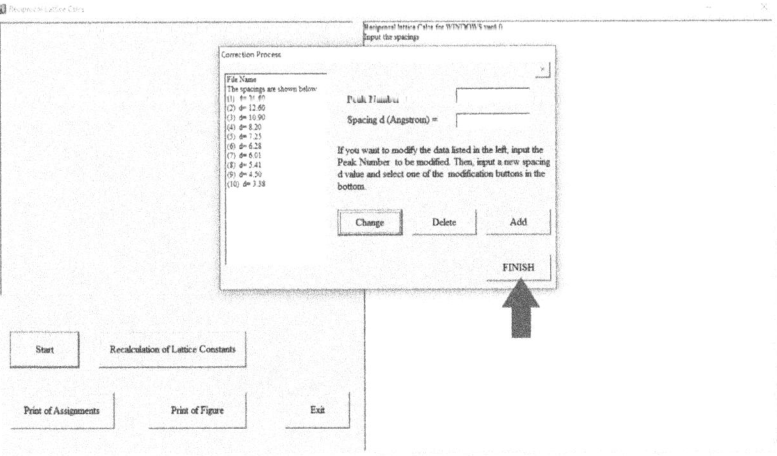

Step (8) Save the data as your convenient name.

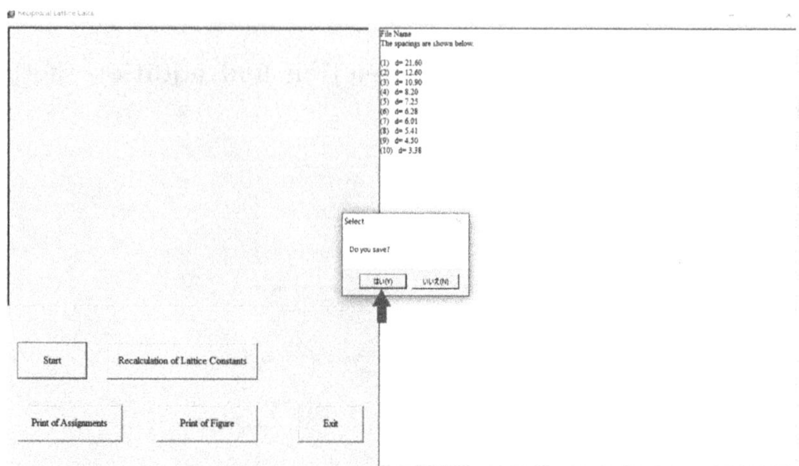

When you are asked the problem of "Do you save?," press the [Yes(Y)] button to save.

Here, the spacing values are saved as the name of $[(C_5O)_4PDP]_2Pd$ at 130C.rcp. Each dataset is stored with the extension ".rcp" after the name. The meaning of rcp is "reciprocal."

Step (9) After saving the data, you will be asked as "Do you calculate the lattice constants?." If you press the [No(N)] button, you can return to the initial "**Application Menu**" screen. If you press [Yes(Y)], you can continue the analysis with this data.

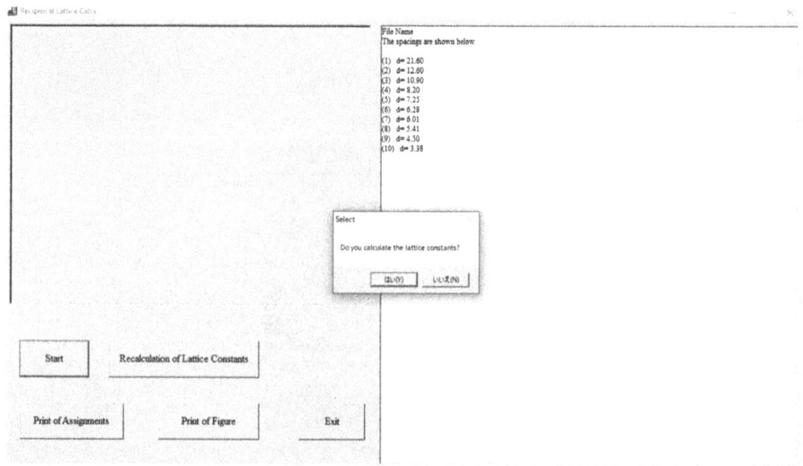

Section 1. Structure Analysis of Columnar Liquid Crystal Phases

2 How to Analyse @M1_Col$_{ho}$

2.1 Data download and estimation of Dimensionality of the mesophase

Step (1) Load the data. Click the [Recip] button.

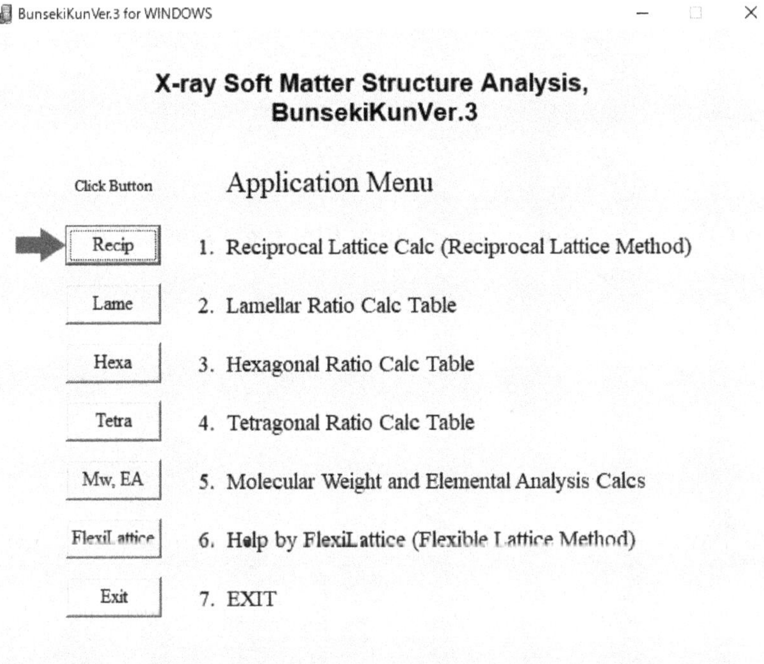

Step (2) Click [Start] button, and then [Read File] button.

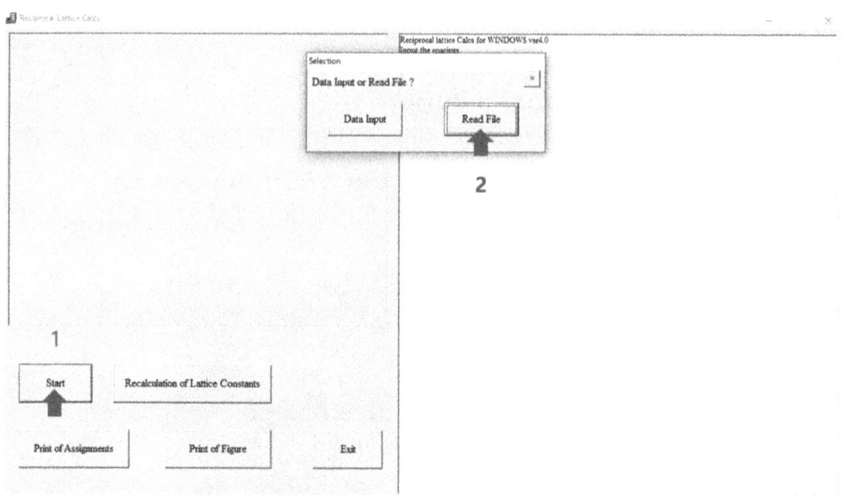

Step (3) Select your desired data file, e.g., @M1, in the RCP extension files.

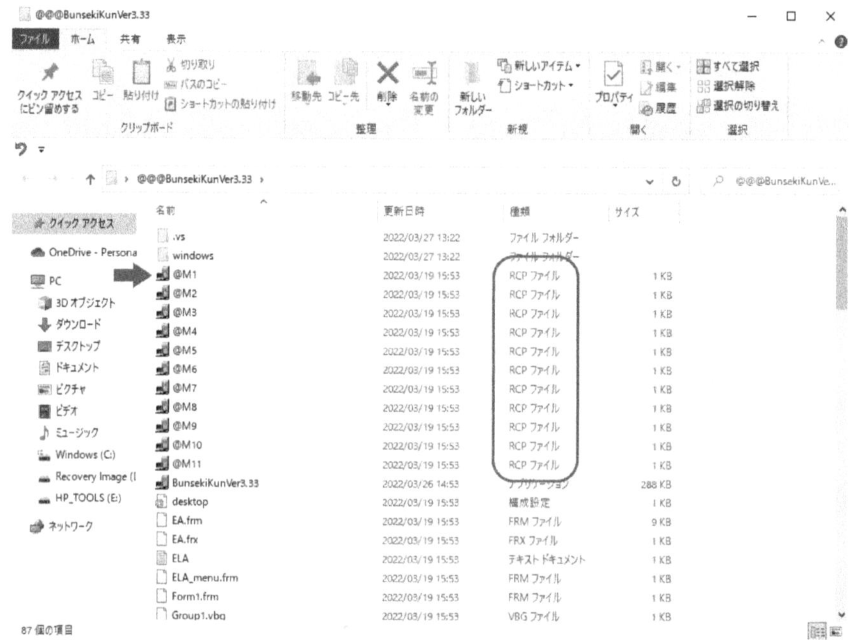

Step (4) If necessary, make correction, change, deletion and addition of the data. Then, click [FINISH] button and [Exit] button.

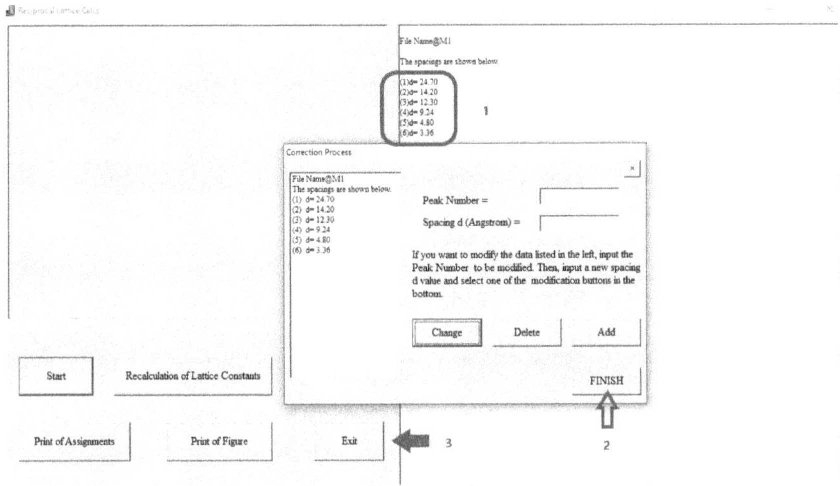

Step (5) Estimation of the dimensionality and the phase structure type of liquid crystal (soft matter) from the downloaded X-ray data.

As already mentioned in the solution for Problem 7, when you look at the X-ray diffraction data, you should use the four articles of "golden rules for liquid crystal structure analysis" in order to estimate the dimensionality and phase structure type of the liquid crystal (soft matter) phase. Roughly judging from Article 2 in the rule to the ratios of the observed spacings, this phase may have a 2D-hexagonal phase. Therefore, here you use the estimation program, [Hexa], to confirm this estimation.

Press the [Hexa] button in the application menu.

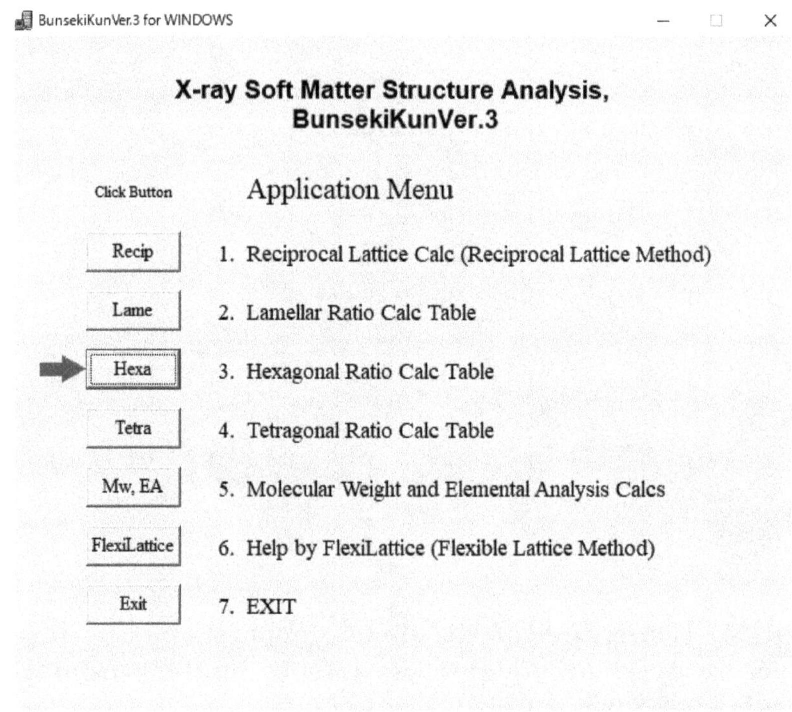

Step (6) When you have pressed the [HEXA] button, a following screen appears.

Click the [Load Data File] button and then select your desired file, @M1 in the data file folder.

Hexagonal Ratio Calc Table (HEXA)
Load Data File
PRINT
EXIT

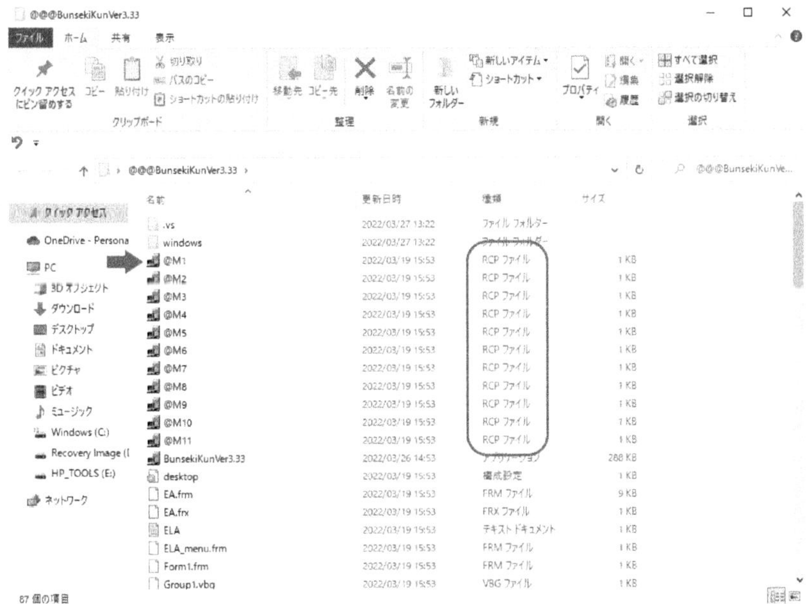
@@@BunsekiKunVer3.33
ファイル
ホーム
共有
表示
クイック アクセスにピン留めする
コピー
貼り付け
切り取り
パスのコピー
ショートカットの貼り付け
クリップボード
移動先
コピー先
削除
名前の変更
整理
新しいフォルダー
新しいアイテム
ショートカット
新規
プロパティ
開く
編集
履歴
すべて選択
選択解除
選択の切り替え
選択
@@@BunsekiKunVer3.33
名前
更新日時
種類
サイズ
クイック アクセス
OneDrive - Persona
PC
3D オブジェクト
ダウンロード
デスクトップ
ドキュメント
ピクチャ
ビデオ
ミュージック
Windows (C:)
HP_TOOLS (E:)
ネットワーク
.vs
windows
@M1
@M2
@M3
@M4
@M5
@M6
@M7
@M8
@M9
@M10
@M11
BunsekiKunVer3.33
desktop
EA.frm
EA.frx
ELA
ELA_menu.frm
Form1.frm
Group1.vbg
RCP ファイル
構成設定
FRM ファイル
FRX ファイル
テキスト ドキュメント
VBG ファイル
87 個の項目

Step (7) When you have selected the data file @M1, the hexagonal ratio calculation table appears as follows.

Hexagonal Ratio Calc Table (HEXA)

Hexagonal Ratio Calc Table (HEXA)

1 : sqr3 : 2 : sqr7

	d	SQR01	SQR03	SQR04	SQR07	SQR09	SQR12	SQR13	SQR16	SQR19
(01)	024.700	028.521	049.400	057.042	075.460	085.563	098.800	102.834	114.084	124.321
(02)	014.200	016.397	028.400	032.793	043.382	049.190	056.800	059.119	065.587	071.472
(03)	012.300	014.203	024.600	028.406	037.577	042.608	049.200	051.209	056.811	061.909
(04)	009.240	010.669	018.480	021.339	028.229	032.008	036.960	038.469	042.678	046.507
(05)	004.800	005.543	009.600	011.085	014.664	016.628	019.200	019.984	022.170	024.159
(06)	003.360	003.880	006.720	007.760	010.265	011.639	013.440	013.989	015.519	016.912

	d	SQR21	SQR25	SQR27	SQR28	SQR31	SQR36	SQR37	SQR39	SQR43
(01)	024.700	130.700	142.606	148.200	150.919	158.799	171.127	173.487	178.114	187.025
(02)	014.200	075.139	081.984	085.200	086.763	091.293	098.380	099.738	102.398	107.521
(03)	012.300	065.085	071.014	073.800	075.154	079.078	085.217	086.392	088.697	093.134
(04)	009.240	048.893	053.347	055.440	056.457	059.405	064.017	064.900	066.631	069.964
(05)	004.800	025.399	027.713	028.800	029.328	030.860	033.255	033.714	034.613	036.345
(06)	003.360	017.779	019.399	020.160	020.530	021.602	023.279	023.600	024.229	025.442

Load Data File PRINT EXIT

As can be seen from this table, the following equation stands as

$$d_1 \times 1 = d_2 \times \sqrt{3} = d_3 \times 2 = d_4 \times \sqrt{7} = 28.4\,\text{Å}$$

This is compatible with the following ratios

$$d_1 : d_2 : d_3 : d_4 = 1 : \frac{1}{\sqrt{3}} : \frac{1}{2} : \frac{1}{\sqrt{7}}$$

Therefore, the ratios of spacing values clearly show that the liquid crystal phase has a two-dimensional hexagonal structure, from Article 2 in the "golden rules for liquid crystal structure analysis".

2.2 Liquid crystal phase structure analysis by Reciprocal Lattice Method

2.2.1 *Data download*

Step (8) Press the [Recip] button in the application selection menu, similarly to **Step (1)**.

Step (9) Click [Start] button, and then [Read File] button, similarly to **Step (2)**.

Step (10) Select your desired data file, e.g., @M1, in the RCP extension files, similarly to **Step (3)**.

Step (11) If necessary, make correction, change, deletion and addition of the data. Then, click [FINISH] button and [Exit] button, similarly to **Step (4)**.

Step (12) When you click [FINISH] button, the following screen appears.

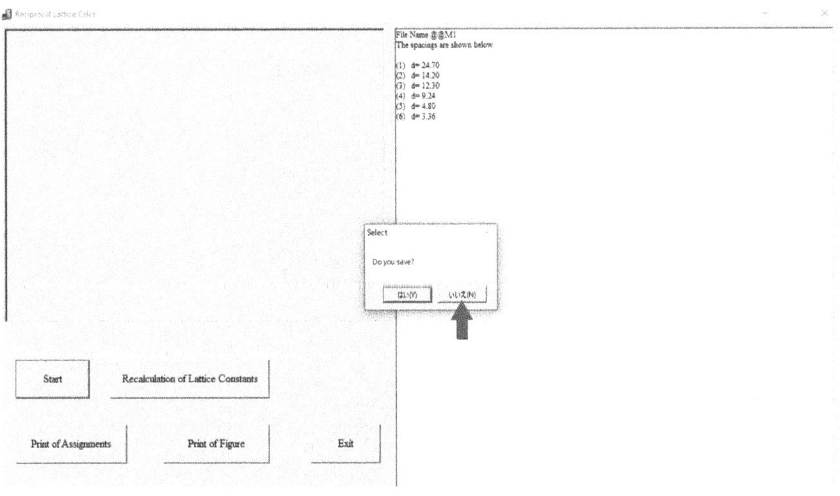

Click [No(N)] button.

2.2.2 *Calculation of the lattice constants*

Step (13) When the following screen appears, click [Yes(Y)] button to calculate the lattice constants.

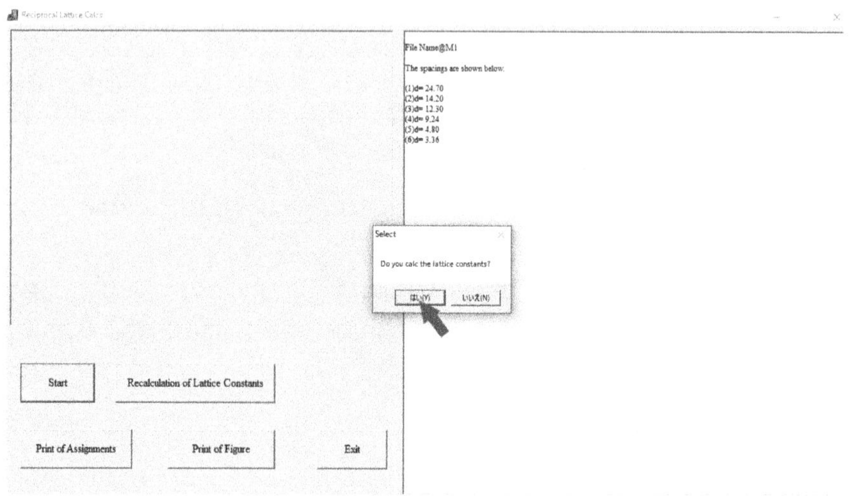

Step (14) Select and click the [2D-hexagonal] button.

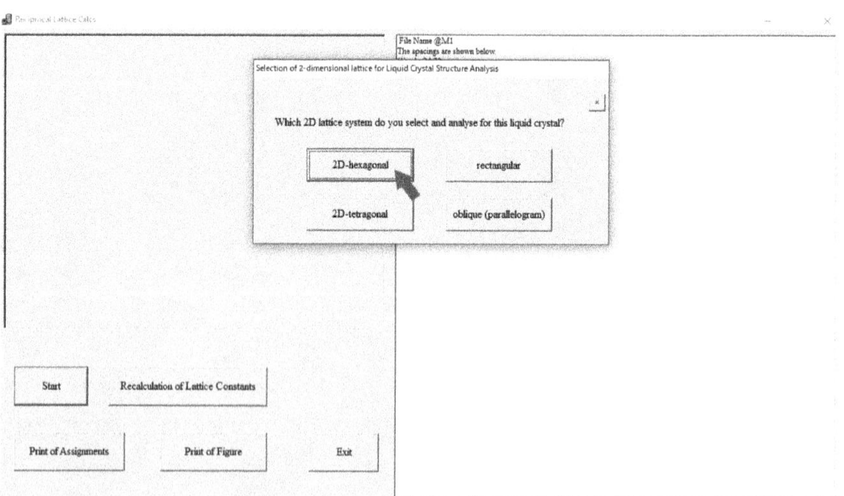

Step (15) When the following screen appears, you select one standard peak for the analysis and input the (hk) value.

Here, you select Peak No. 1 assumed as the standard peak with (10). Accordingly, input the values in the blanks indicated with red arrows.

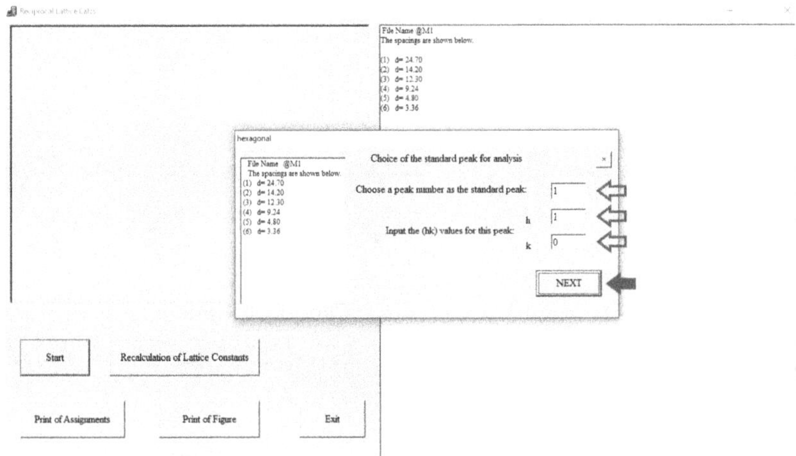

For these inputs, use the Tab key on the keyboard to move among the blanks. Since it is a toggle, if you press the Tab key it many times, it will return to the original blank. When you complete the inputs, press the [NEXT] button.

Step (16) Input the numbers, h and k, for the reciprocal lattice plane divisions. If completed, click the [NEXT] button.

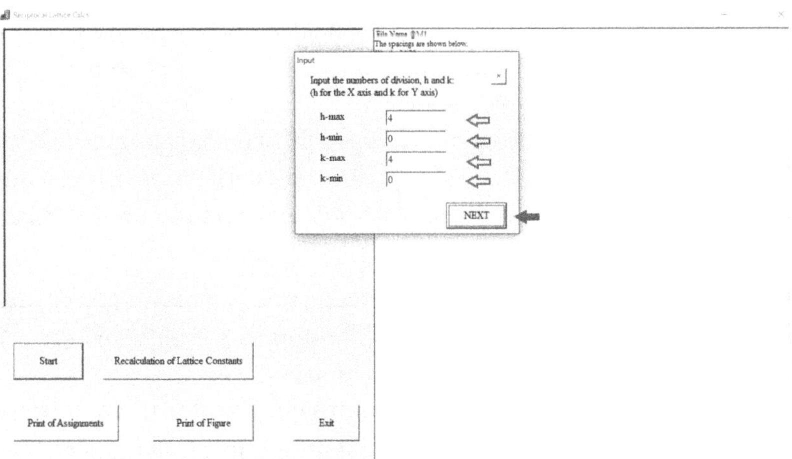

Step (17) When you press the [NEXT] button in the previous (16) step, the Debye-Scherrer rings appear of the X-ray reflections as the quarter circles on the two-dimensional hexagonal reciprocal lattice plane shown below.

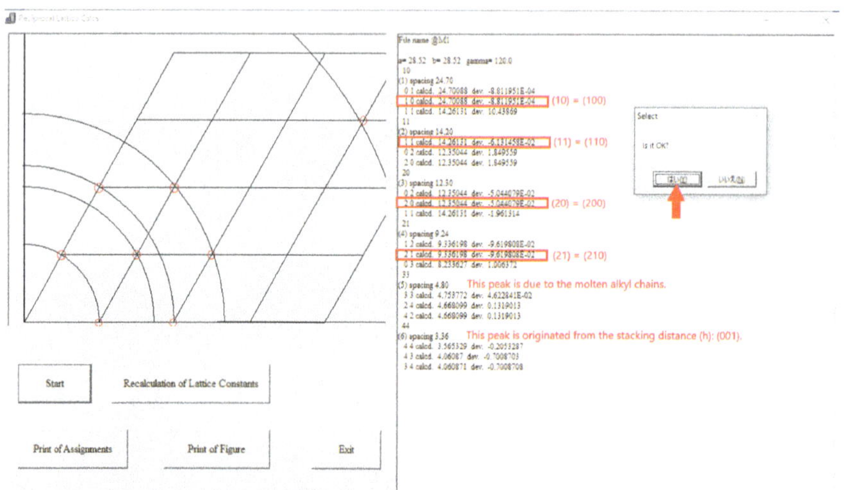

On this reciprocal lattice plane, it can be seen that the Debye-Scherrer rings of Peak Nos. 1 to 4 agree well with the lattice intersection points. When you read out the intersection (hk) points, you can index these four reflections as (10), (11), (20) and (21). Furthermore, as can be seen in this figure, Peak No. 5 is a very broad and large peak, corresponding to the average distance between the fused alkyl groups. Peak No. 6 located at d = 3.36 Å corresponds to the (001) reflection of the stacking distance h between discotic molecules in the column.

From the above-mentioned reciprocal lattice analysis method, the liquid crystalline phase of @M1.rcp can be identified as a hexagonal ordered columnar (Col_{ho}) phase with lattice constants $a = 28.52$ Å and $h = 3.38$ Å.

Step (18) If you are satisfied with the lattice calculation, you press [Yes(Y)] on the screen in the above (17) step. A following screen will appear for verification by Z value calculation.

If the number of molecules in a unit cell of Col_{ho} phase is set as Z, the Z value should be equal to 1. Therefore, if the identification of this liquid crystal phase is correct, the Z value calculated from the lattice constant must be an integer value of 1. If it is not an integer, such as 0.5 or 1.5, the present identification is wrong. Therefore, it can be verified whether this identification is correct or not, by using the Z-value calculation.

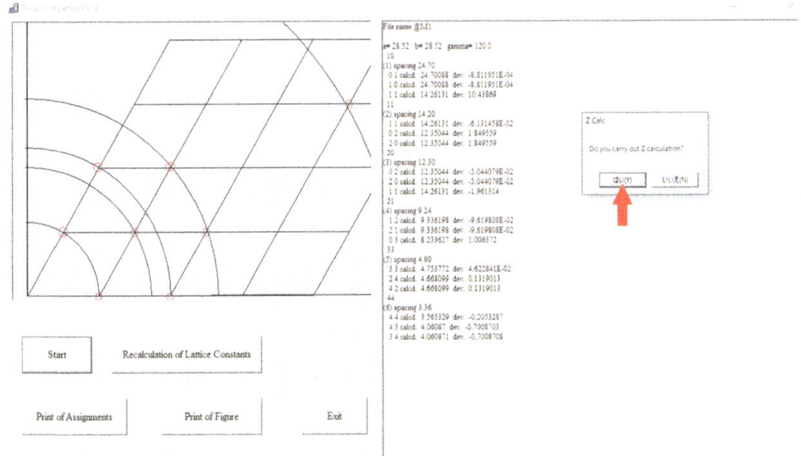

Step (19) For the question [Do you carry out Z calculation?], click the [Yes(Y)] button.

2.2.3 *Verification by Z value calculation*

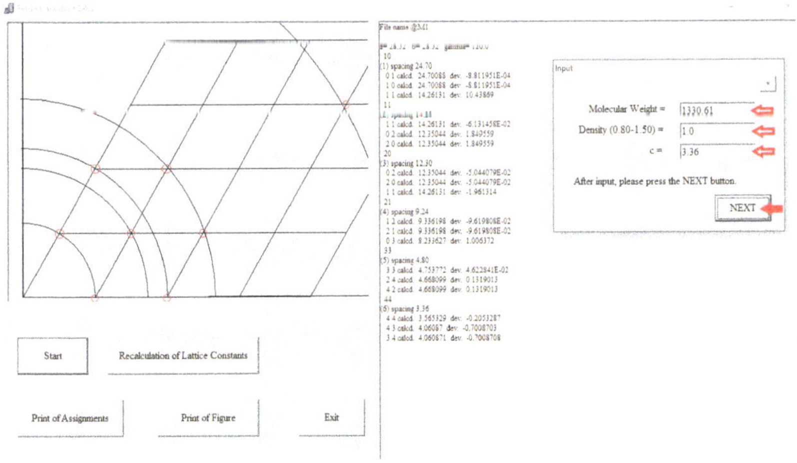

Step (20) Into the blanks in this screen appeared, enter the values of molecular weight$, density and stacking distance h (= inter-disk distance of (001) reflection), and then click the [NEXT] button.

If there is no actually observed value for the density, enter an assumed value within the range of 0.80 to 1.50 g/cm^3 that most liquid crystal phases exhibit. Experimentally, the density of liquid crystals can be measured by the dilatometry method, but it is very time-consuming and troublesome to measure the density while applying temperature. Since the density is used here only for verification, it is sufficient that you use an assumed value in the range of 0.80 to 1.50 g/cm^3. Generally speaking, the derivatives with short alkyl groups have high densities, and the derivatives with long alkyl groups and at high temperatures tend to show densities of 1.0 or less. Therefore, when the density is assumed to be 1.0 in this case.

Accordingly, you can obtain as Z = 1.07 ≅ 1.0, as shown in the bottom. Thus, the calculated Z value is integer 1, so that this identification of $\mathrm{Col_{ho}}$ is consistently correct.

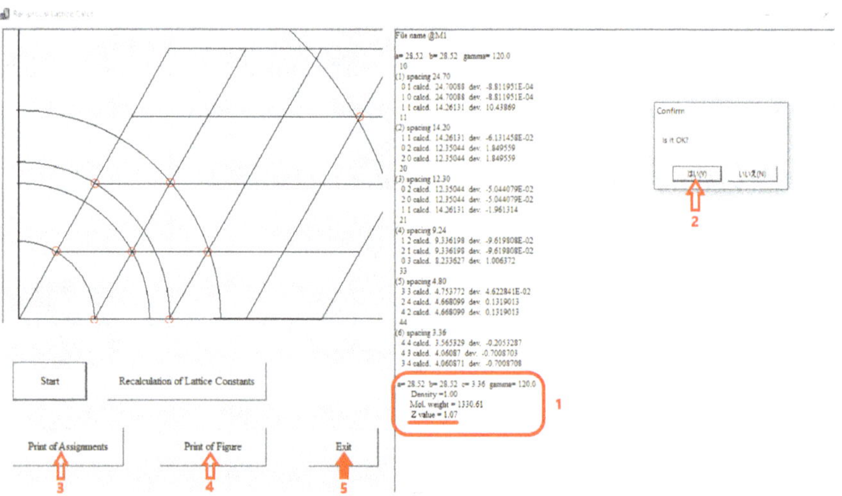

If you are OK for the confirmation, press the [Yes(Y)] button. If necessary, press [Print of Figure] and [Print of Assignments] buttons.

The liquid crystal structure analysis results are summarized and tabulated as follows.

Table 4. X-ray data of @M1 phase and the indexation of the reflections.

Entry No.	M1		
Reference	1, Ni(12,1̃)		
Mw	Mw=1330.61		
Temperature	170°C		
Mesophase	Col_{ho}		
Lattice constants	a =28.5, h = 3.36; Z = 1.07 for ρ = 1.00		
Peak No.	$d_{obs.}$	$d_{calcd.}$	*(hkl)*
1	24.7	24.7	(1 0 0)
2	14.2	14.3	(1 1 0)
3	12.3	12.4	(2 0 0)
4	9.24	9.34	(2 1 0)
5	*ca.* 4.8	–	#
6	3.36	3.36	*h*: (0 0 1)

$Note: The molecular weight can be calculated by selecting the subprogram "5. Molecular Weight and Elemental Analysis Calcs" in the initial screen of Application Menu. When you use this subprogram, be sure to enter the integer 1 in the molecular formula even where there is only one element, as follows:

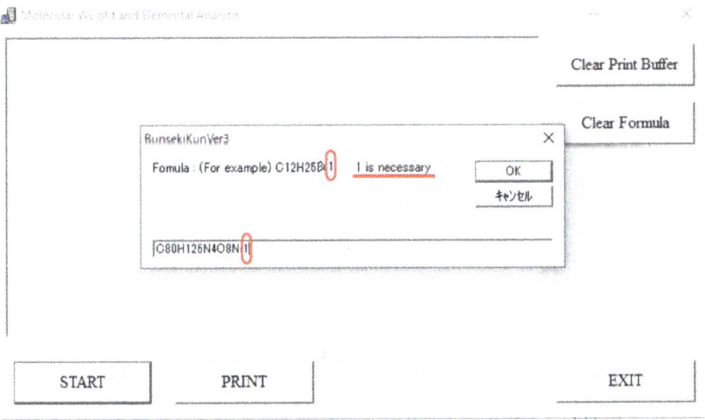

3 How to Analyse @M4_Col$_{rd}$(C2/m)

3.1 Loading X-ray diffraction data

Step (1) In the initial screen [Application Menu] of the Bunseki-kun Ver.3, press sequentially the buttons, [Recip] → [Start] → [Read File], in order to download the X-ray data of @M4.

Step (2) Look at the X-ray data and estimate what kind of dimensionality the liquid crystal has. Therefore, the golden rules of liquid crystal structure analysis are sequentially examined from Article 1 to Article 4.

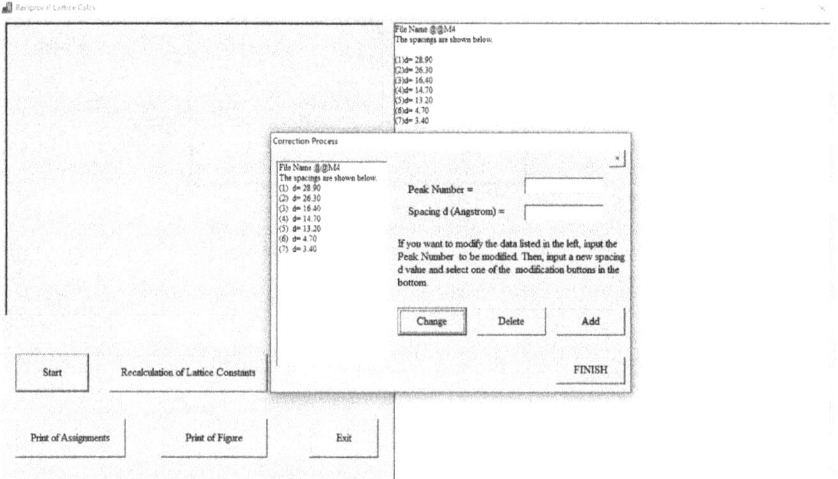

A quick look at these values suggests two series of lamellar structures may exist.

$$28.9\text{: } 14.7 = 1\text{: } 1/2$$

$$26.3\text{: } 13.2 = 1\text{: } 1/2$$

Accordingly, this seems to be a rectangular phase from Article 4 in the golden rule of liquid crystal structure analysis.

Moreover, two peaks, No. 1 and No. 2, appear close to each other in the low angle region, which is one of the characteristics of the rectangular phase.

Therefore, it can be assumed to be a liquid crystal phase having a rectangular lattice.

3.2 Use subprogram [Lame] to confirm that two series of lamellae exist in this phase

Step (3) Press the [Lame] button and then click the [Load Data File] button. Select the target file, @M4 to download.

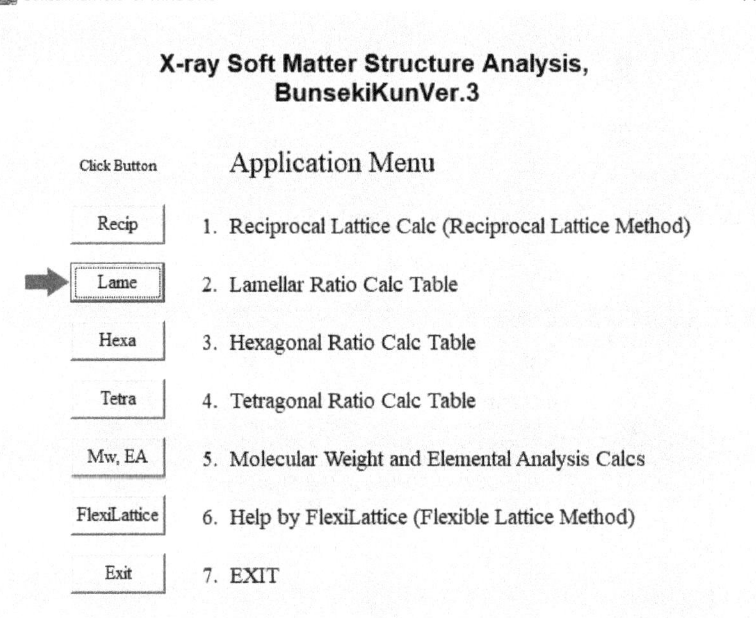

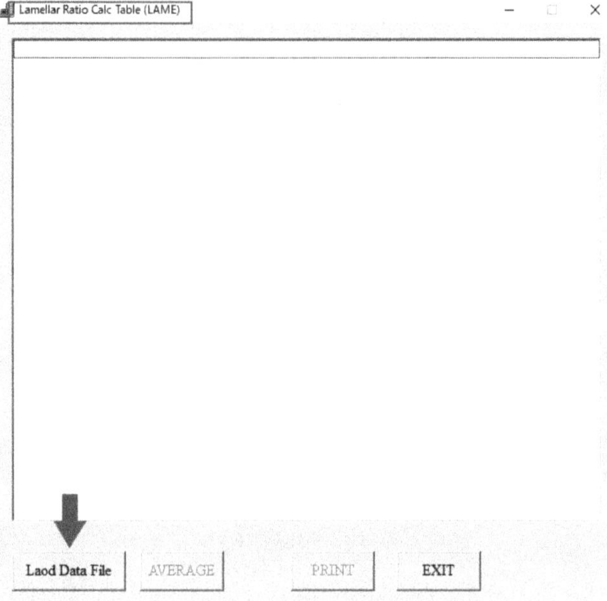

The following lamellar ratio calculation table appears.

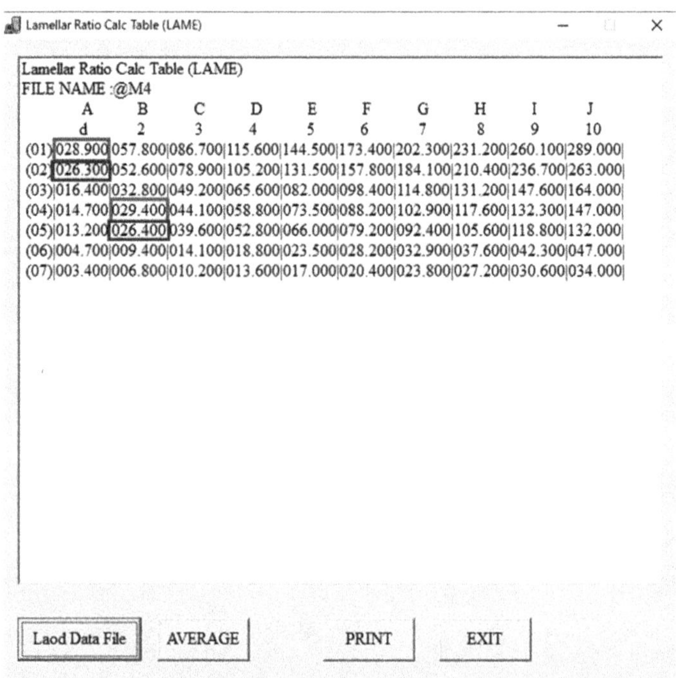

	A	B	C	D	E	F	G	H	I	J
	d	2	3	4	5	6	7	8	9	10
(01)	028.900	057.800	086.700	115.600	144.500	173.400	202.300	231.200	260.100	289.000
(02)	026.300	052.600	078.900	105.200	131.500	157.800	184.100	210.400	236.700	263.000
(03)	016.400	032.800	049.200	065.600	082.000	098.400	114.800	131.200	147.600	164.000
(04)	014.700	029.400	044.100	058.800	073.500	088.200	102.900	117.600	132.300	147.000
(05)	013.200	026.400	039.600	052.800	066.000	079.200	092.400	105.600	118.800	132.000
(06)	004.700	009.400	014.100	018.800	023.500	028.200	032.900	037.600	042.300	047.000
(07)	003.400	006.800	010.200	013.600	017.000	020.400	023.800	027.200	030.600	034.000

As can be seen from this table, the following equation stands as

$$d_1 \times 1 = d_4 \times 2 = 29.2\,\text{Å}$$

(Use [AVERAGE] program.)

$$d_2 \times 1 = d_5 \times 2 = 26.4\,\text{Å}$$

(Use [AVERAGE] program.)

They are compatible with the following ratios.

$$d_1 : d_4 = 1 : \frac{1}{2}$$

$$d_2 : d_5 = 1 : \frac{1}{2}$$

Thus, the ratios of spacing values clearly show that the liquid crystal phase has two series of lamellar structures. Therefore, this phase can be judged as a rectangular phase from Article 4 in the "golden rules for liquid crystal structure analysis."

If necessary, press the [PRINT] button. Then, press the [EXIT] button.

3.3 Liquid crystal structure analysis by the method using the standard peaks

Step (4) Loading X-ray diffraction data.

When pressed the [EXIT] button, you return to the initial screen [Application Menu]. Press sequentlally the buttons, [Recip] → [Start] → [Read File] → [FINISH] → [No(N)], in order to download the @M4's X-ray data. A following screen appears and you are asked as "Do you calculate the lattice constants?". Press [Yes(Y)] and then [rectangular].

Step (5) Then, the following screen appears. Peak Nos. 1 and 2 are assumed as the standard reflection peaks from the (200) and (110) planes, respectively.

Input the (hk) values as follows:

i.e.,

Peak No. 1 = the first standard peak, assume (hk) = (20)

Peak No. 2 = the second standard peak, assume (hk) = (11)

and then press the [NEXT] button.

Step (6) When the following screen appears, input the tentative numbers of division range for h and k, and then press the [NEXT] button.

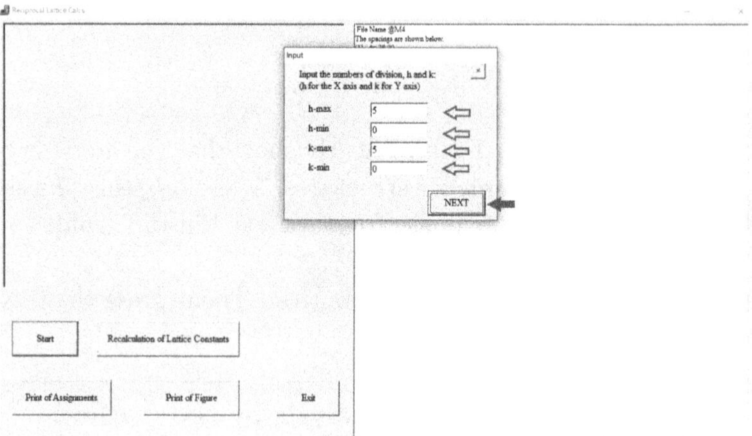

Step (7) Then a reciprocal lattice plane corresponding to the two-dimensional rectangular lattice and the quarter circles corresponding to the Debye-Scherrer rings appear.

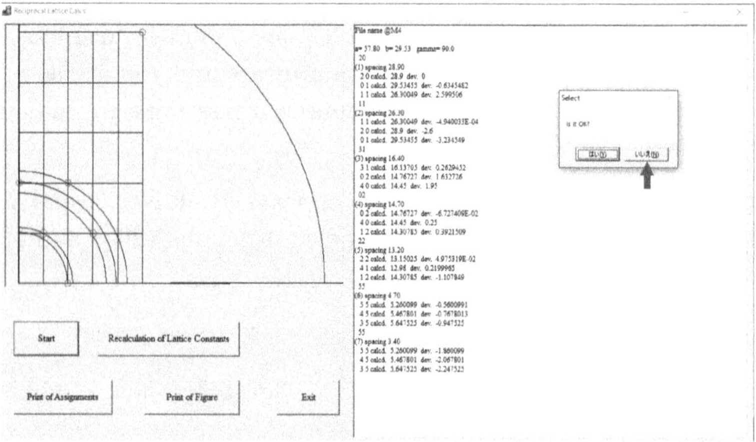

Step (8) Since the tentative division range of the reciprocal lattice plane is not appropriate, you input again more appropriate division range. Press the [No(N)] button and then the [Input new values of h and k] button, as follows:

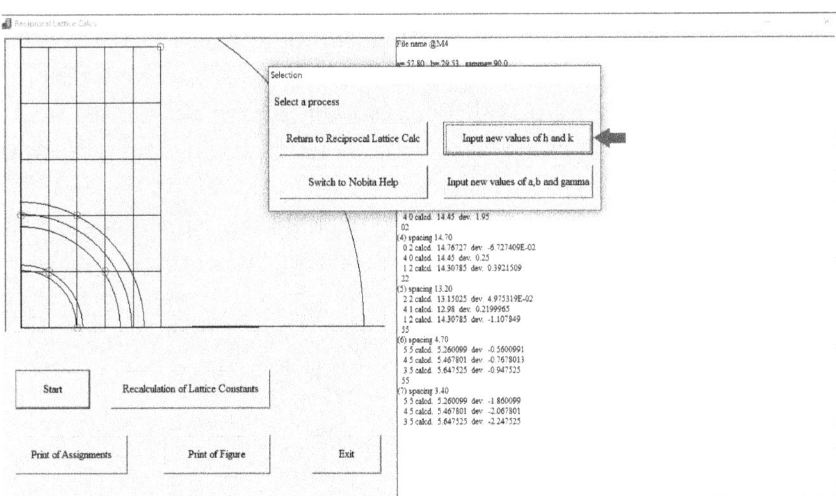

Step (9) Input more appropriate numbers of division range, h and k, and then press the [NEXT] button.

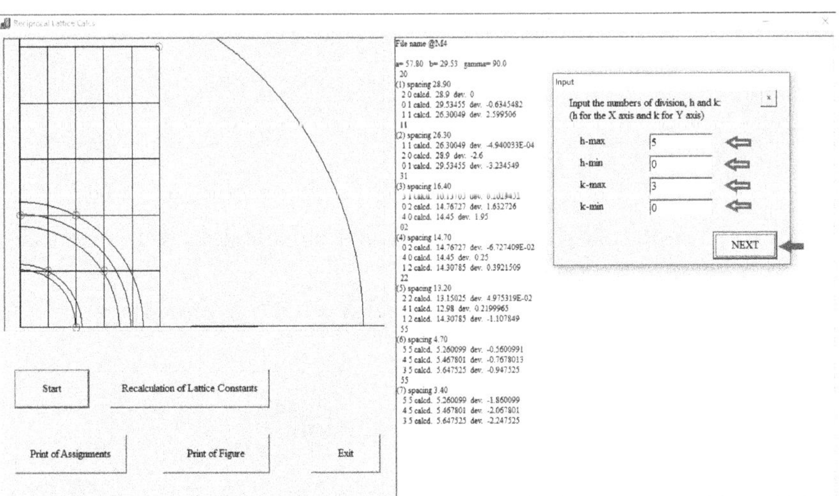

Step (10) Calculation of the lattice constants and indexation of the reflection peaks by Reciprocal Lattice Method.

A following reciprocal lattice plane with more appropriate numbers of division range is shown. Each of the intersection points with the quarter circles is highlighted with small red circle.

When you read out the intersection (hk) points, you can index the first five reflections as (20), (11), (31), (02) and (22). Furthermore, the latter two peaks are broad. Peak No. 5 is due to the molten alkyl chains. Peak No. 6 located at d = *ca.*3.4 Å is due to the stacking distance h of the (001) reflection.

Thus, the liquid crystalline phase of @M4.rcp can be identified as a rectangular columnar (Col_r) phase with lattice constants a = 57.80 Å and b = 29.53 Å.

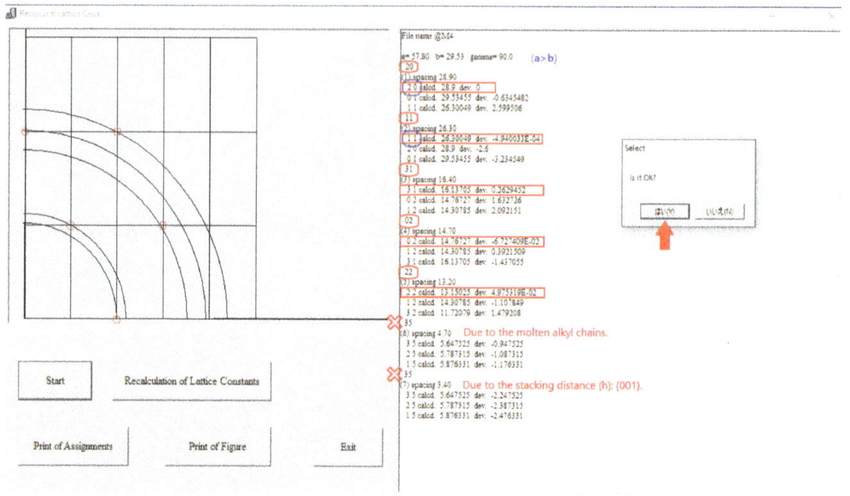

If you are satisfied with this lattice calculation, you press the [Yes(Y)] button.

Note: To become a > b, assume d_{20} = 28.9 Å and d_{11} = 26.3 Å in this case.

3.4 Verification by Z value calculation

Step (11) When you press the [Yes(Y)] button in the above (9) step, a following screen appears for verification by Z value calculation.

For the question [Do you carry out Z calculation?], click the [Yes(Y)] button.

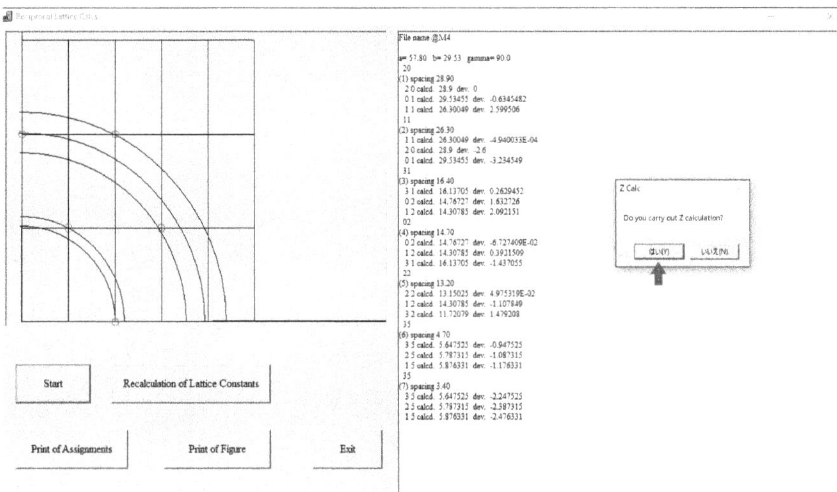

Step (12) Into the blanks in a screen appeared, enter the values of molecular weight, density and c (= stacking distance h)$^{\$}$, and then click the [NEXT] button.

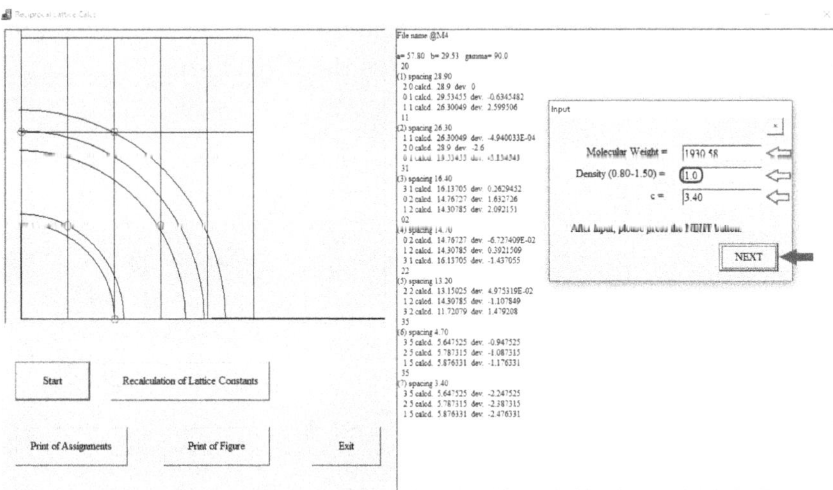

[$]Note: Although Peak No. 6 located at d = *ca.* 3.4 Å is not sharp but broad, it corresponds to the stacking distance $h(= c)$.

Step (13) Evaluation of the obtained Z value.

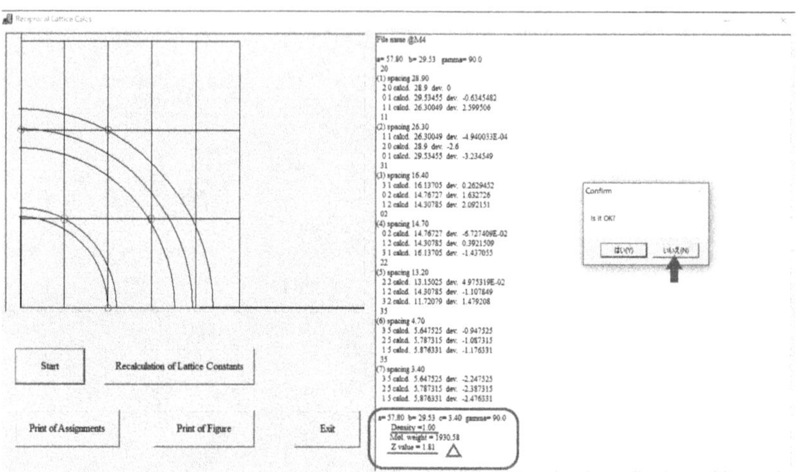

Thus, you obtain as Z = 1.81, as shown in the bottom. The calculated Z value is a little bit far from integer 2, so that you calculate again using another assumed density.

Step (14) Recalculation of the Z value using another density. Input another density of 1.10, and then press the [NEXT] button.

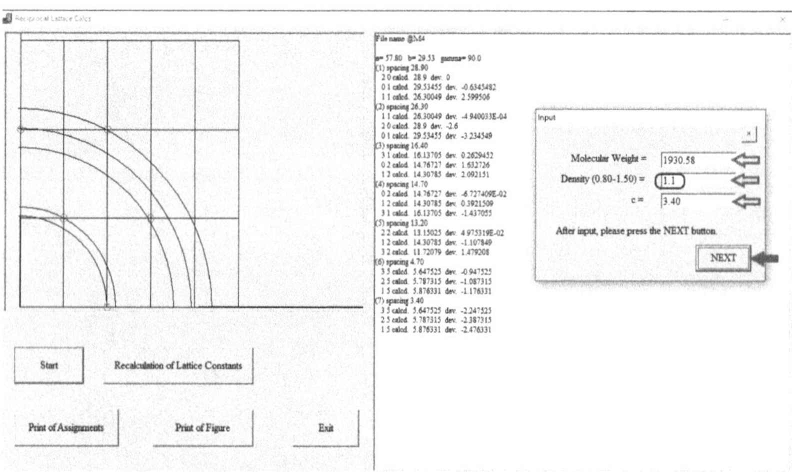

Step (15) Re-evaluation of the obtained Z value

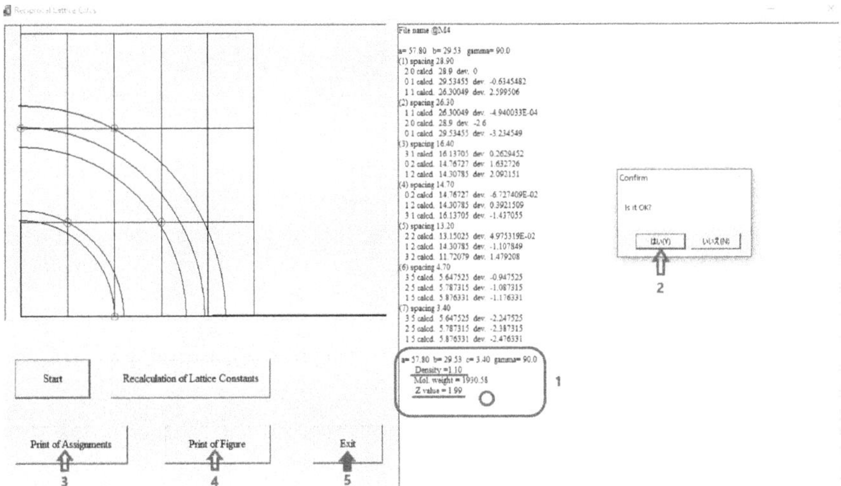

Thus, you obtain as Z $= 1.99 \cong 2.0$, as shown in the bottom. Therefore, the calculated Z value is integer 2, so that this identification of Col_r is consistently correct.

Step (16) If you are OK for the confirmation, press the [Yes(Y)] button. If necessary, press the [Print of Figure] and [Print of Assignments] buttons. Finally, press the [EXIT] button.

However, four symmetries are known so far for two-dimensional rectangular lattices, as shown in Figure 3. Among them, the symmetries with Z = 2 are C2/m and $P2_1/a$. Therefore, the final structure cannot be determined unless this symmetry is determined. Each symmetry has a characteristic extinction rule. Accordingly, you will further estimate the liquid crystalline phase structure from the extinction rules shown in Figure 3.

Step (17) Estimation from the extinction rules.

From Figure 3, the extinctions are

For C2/m symmetry, hk: $h + k = 2n + 1$

For $P2_1/a$ symmetry, $h0$: $h = 2n + 1$; $0k$: $k = 2n + 1$

When you look at the indices circled with red ink in **Step (10)**, there are no reflections when $h +$ k I $= 2n + 1$. Therefore, this liquid

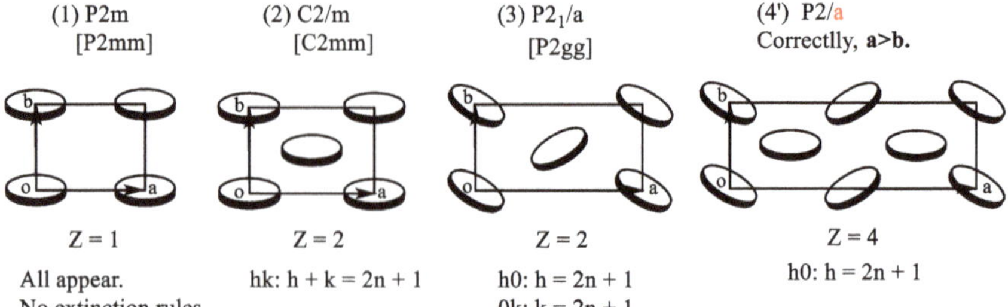

Figure 3. Summary of the Z values, the extinction rules and the symmetries of four rectangular lattices found to date. The length of the lattice constant a should be taken longer than b from the precise rule for crystallography.

crystal phase can be identified as a rectangular columnar phase Col_r having a C2/m symmetry.

The liquid crystal structure analysis results are summarized and tabulated as follows.

Table 5. X-ray data of @M4 and the indexation of the reflections.

Entry No.	M4		
Reference	p30, (C_{12}OPzCu		
Mw	Mw=1930.58		
Temperature	150°C		
Mesophase	Col_{rd}(C2/m)		
Lattice constants	a = 57.8, b = 29.5, h = 3.4; Z = 1.99 for ρ = 1.10		
Peak No.	$d_{obs.}$	$d_{calcd.}$	(*hkl*)
1	28.9	28.9	(2 0 0)
2	26.3	26.3	(1 1 0)
3	16.4	16.1	(3 1 0)
4	14.7	14.8	(0 2 0)
5	13.2	13.2	(2 2 0)
6	*ca.* 4.7	–	#
7	*ca.* 3.4	*ca.* 3.4	*h*: (0 0 1)

$Note: Although Peak No. 6 located at d = *ca.* 3.4 Å surely corresponds to the stacking distance $h(= c)$, it is not sharp but broad. Therefore, this phase is identified not $\mathrm{Col_{ro}}$(C2/m) but $\mathrm{Col_{rd}}$(C2/m). The difference between the ordered phase and the disordered phase is not so strict for the definition.

4 How to Analyse @M5_$\mathbf{Col_{rd}(P2_1/a)}$

When you assume d_{11} = 32.0 Å and d_{20} = 28.3 Å in the observed values of the @M5 phase listed in the table below, the calculation is carried out in the same way mentioned in the above @M4 phase, and a = 56.6 Å and b = 38.8 Å are obtained. The reciprocal lattice and the quarter circles of the reverse values of spacings are drawn as follows:

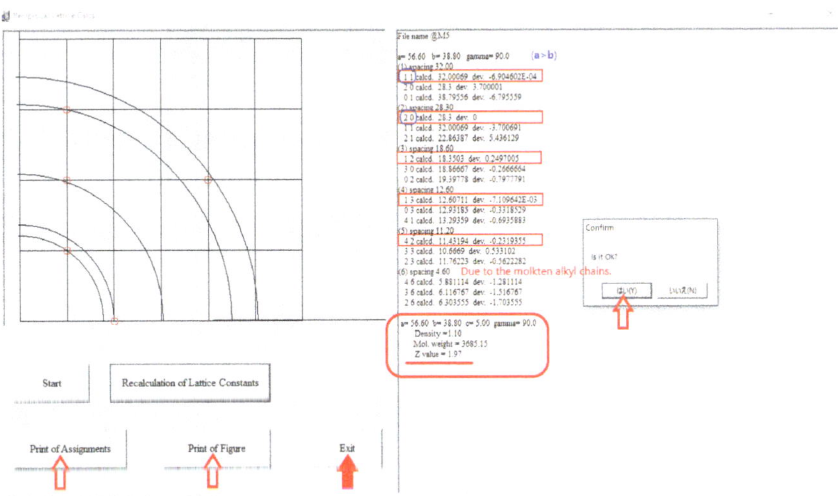

By reading the intersection points of the reciprocal lattice and the quarter circles, the reflection lines at 32.0, 28.3, 18.6 Å, 12.6 Å and 11.2 Å can be indexed as (11), (20), (12), (13) and (42) respectively. From the extinction rules in Figure 3, it can be concluded that this rectangular lattice has a symmetry of $\mathrm{P2_1/a}$. As can be seen from the figure and table, d_{001} does not appear for @M5. Assuming c = 5.0 Å, Z = 1.97 ≈ 2 is obtained, which is consistent with the theoretical value Z = 2 for a rectangular lattice having a $P2_1/a$ symmetry. Therefore, this @M5 phase can be identified as the Col_{rd} ($P2_1/a$) phase.

Note: To become a > b, assume d_{11} = 32.0 Å and d_{20} = 28.3 Å in this case.

The liquid crystal structure analysis results are summarized and tabulated as follows.

Table 6. X-ray data of @M5 and the indexation of the reflections.

Entry No.	M5		
Reference	p60, C_{10}-Cu		
Mw	Mw=3685.15		
Temperature	100°C		
Mesophase	$Col_{rd}(P2_1/a)$		
Lattice constants	a = 56.6, b = 38.8, h = 5.0; Z = 2.0 for ρ = 1.1		
Peak No.	$d_{obs.}$	$d_{calcd.}$	(*hkl*)
1	32.0	32.0	(1 1 0)
2	28.3	28.3	(2 0 0)
3	18.6	18.4	(1 2 0)
4	12.6	12.9	(1 3 0)
5	11.2	11.4	(4 2 0)
6	*ca.* 4.6	–	#

5 How to Analyse @M6_Col_{rd}(P2/a)

5.1 Reading the X-ray data file and calculation of the lattice constants.

Step (1) In the initial screen [Application Menu] of the Bunseki-kun Ver.3, press sequentially the buttons, [Recip] → [Start] → [Read File], in order to download the X-ray data of @M6.

Step (2) Look at the X-ray data and estimate what kind of dimensionality the liquid crystal has. Judging from Article 1 to

Article 4 in the golden rules of liquid crystal structure analysis like as the previous examples, this phase may have a rectangular lattice.

Step (3) Accordingly, press sequentially the buttons, [FINISH] → [No(N)], and you are asked as "Do you calculate the lattice constants?". Press [Yes(Y)] and then [rectangular].

Step (4) Then, the following screen appears. Peak Nos. 1 and 2 are assumed as the standard reflection peaks from the (200) and (110) planes, respectively.

Input the (hk) values as follows:

i.e.,

Peak No. 1 = the first standard peak, assume (hk) = (11)

Peak No. 2 = the second standard peak, assume (hk) = (20)

and then press the [NEXT] button.

Step (5) When the following screen appears, input the appropriate numbers of division range for h and k, *i.e.*, input h = (8, 0) and k = (8, 0).

Then press the [NEXT] button. The following a screen appears.

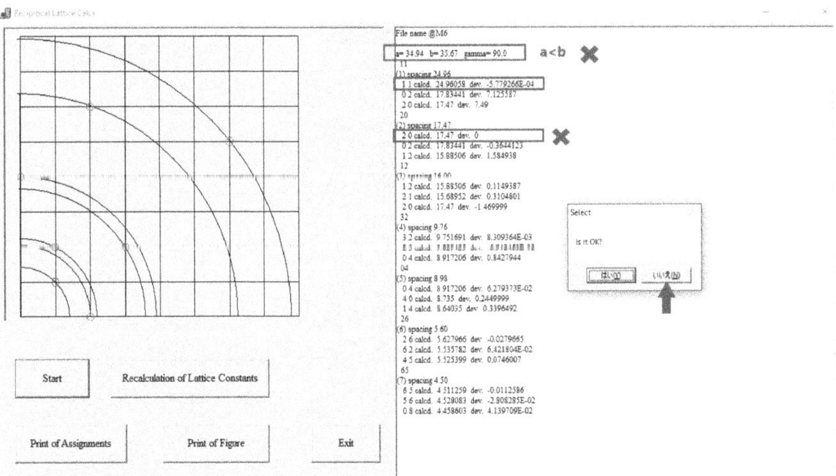

Step (6) When you look at the right table in this screen, it shows a = 34.94, b = 35.67 (a < b). These lattice constants are out of the rule a > b for the rectangular lattices.

Therefore, in order to accord with the rule, press the No(N) button for the question "Is it OK?."

Then, press sequentially the buttons, [Return to Reciprocal lattice] → [Recalculation of Lattice Constants] → [Yes(Y)] → [rectangular]. The following screen appears.

Step (7) Input the (hk) values again as follows:

Peak No. 1 = the first standard peak, assume (hk) = (11).

Peak No. 2 = the second standard peak, assume (hk) = (02)

You obtain a = 35.67, b = 34.94 in accordance with the rule a > b.

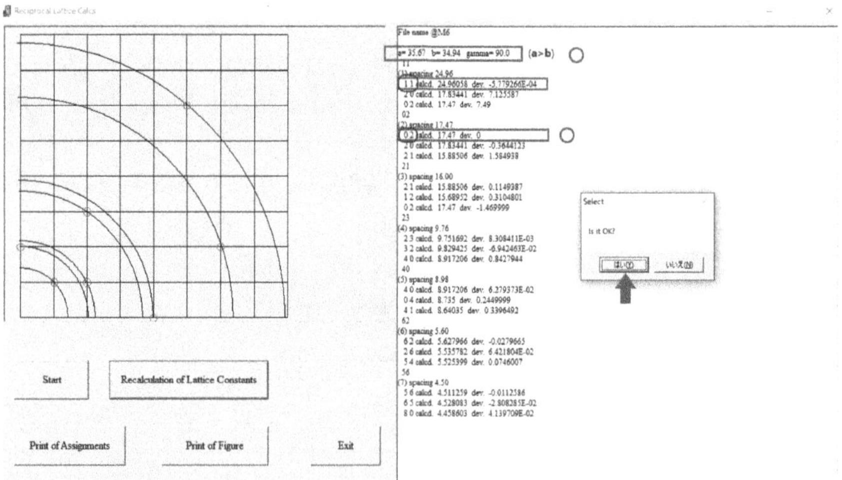

On this reciprocal lattice plane, it can be seen that the Debye-Scherrer rings of Peak Nos. 1 to 5 agree well with the lattice intersection points. When you read out the intersection (*hk*) points, you can index these five reflections as (11), (02), (21), (23) and (40). Furthermore, as Peak Nos. 6 and 7 are broad peaks, corresponding to the (001) reflection of the stacking distance h between discotic molecules in the column, and the average distance between the fused alkyl groups, respectively.

From the reciprocal lattice analysis method, the liquid crystalline phase of @M6 can be thus identified as a rectangular disordered columnar ($\mathrm{Col_{rd}}$) phase with lattice constants a = 35.7 Å and b = 34.9Å and h = *ca.*5.6 Å.

The Z value (= the number of molecules in a unit cell) should be equal to an integer of 1, 2 or 4, as can be seen from Figure 3. Depending on the integer, you can identify the lattice symmetry of this rectangular phase. Hence, you can determine the final lattice structure by using the Z-value calculation.

Step (8) Therefore, input the molecular weight, an assumed density and the c value in the inset table.

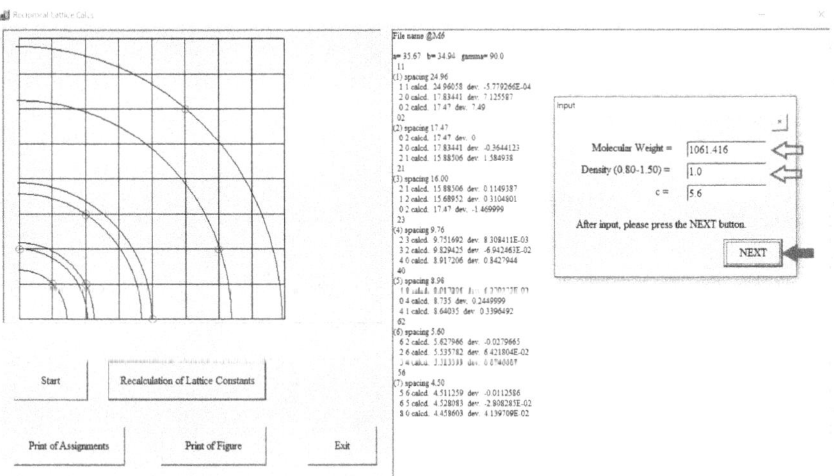

As the result, you obtain Z = 3.96 ≅ 4.0. As shown in Figure 3, four symmetries are known so far for two-dimensional rectangular lattices, and each symmetry has a characteristic extinction rule. Among them, the symmetry with Z = 4 is P2/a. Hereupon, you will further confirm the symmetry from the extinction rules shown in Figure 3.

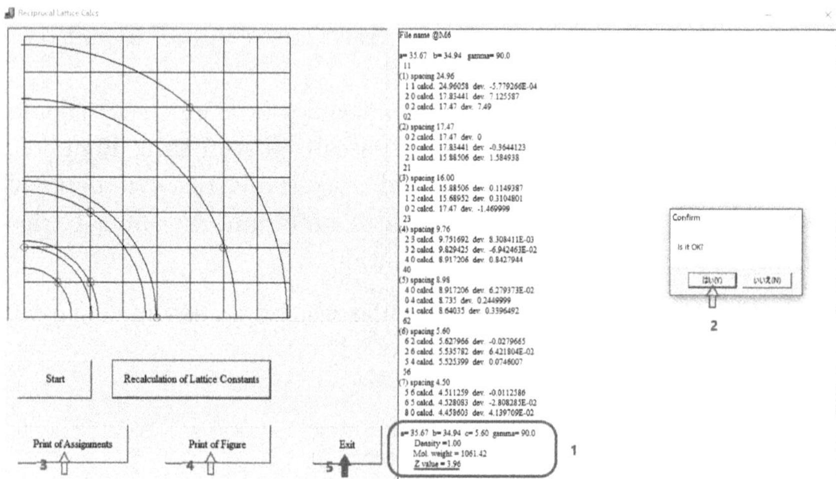

Accordingly, you can obtain as Z = 3.96 ≅ 4.0, as shown in the bottom. Thus, the calculated Z value is integer 4, so that this identification of $\mathrm{Col_{rd}}$ is consistently correct.

If you are OK for the confirmation, press the [Yes(Y)] button. If necessary, press the [Print of Figure] and [Print of Assignments] buttons. Finally, press the [EXIT] button.

Step (9) Estimation from the extinction rules.
However, four symmetries are known so far for two-dimensional rectangular lattices, as shown in Figure 3. Among them, the symmetries with Z = 4 is P2/a. The symmetry has a characteristic

extinction rule. Therefore, the final structure should be confirmed using the extinction rule.

From Figure 3, the extinction for P2/a symmetry is $h0$: h = 2n + 1.

When you look at the indices circled with red ink in **Step (7)**, there are no reflections when $h0$: h = 2n + 1. Therefore, this liquid crystal phase can be consistently identified as a rectangular columnar phase Col_r having a P2/a symmetry.

The liquid crystal structure analysis results are summarized and tabulated as follows.

Table 7. X-ray data of @M6 and the indexation of the reflections.

Entry No.	M6		
Reference	Billard, RHO		
Mw	Mw=1061.416		
Temperature	T in Dc		
Mesophase	Col_{rd}(P2/a)		
Lattice constants	a = 35.7, b = 34.9; Z = 4.0 for ρ = 1.0		
Peak No.	$d_{obs.}$	$d_{calcd.}$	(*hkl*)
1	25.0	25.0	(1 1 0)
2	17.5	17.5	(0 2 0)
3	16.0	15.9	(2 1 0)
4	9.76	9.75	(2 3 0)
5	8.98	8.92	(4 0 0)
6	*ca.* 5.6	broad	h_1: (0 0 1)
7	*ca.* 4.5	–	#

J. Billard, et al., *Mol. Cryst. Liq. Cryst.*, **66**, 115-122(1981).

6 How to Analyse @M7_Col$_{rd}$(P2m)

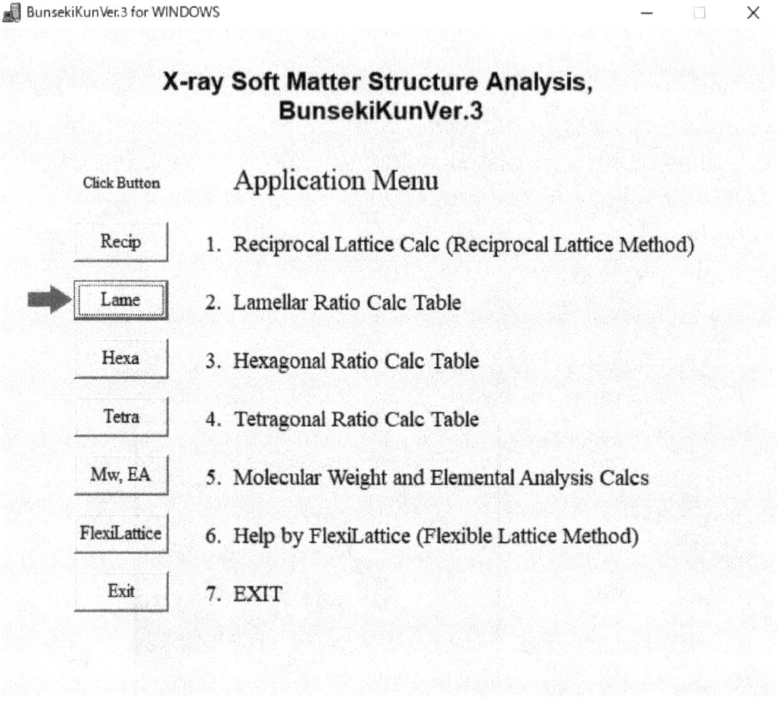

6.1 Loading X-ray diffraction data

Step (1) In the initial screen [Application Menu] of the Bunseki-kun Ver.3, press sequentially the buttons, [Recip] → [Start] → [Read File], in order to download the X-ray data of @M7.

Step (2) Look at the X-ray data and estimate what kind of dimensionality the liquid crystal has. Judging from Article 1 to Article 4 in the golden rules of liquid crystal structure analysis like as the previous examples, this phase may have a rectangular lattice.

6.2 Estimation of the dimensionality and the phase structure type of liquid crystal

Step (3) Accordingly, press sequentially the buttons, [X] in the right up → [EXIT], to return to the initial screen [Application Menu]. Press the [Lame] button to calculate the lamellar ratios.

Step (4) When appeared the left screen, press [Read File]. In the appeared data folder, select the X-ray data of @M7 and then press the [Open(O)] button.

Step (5) When appeared the left Lamellar Ratio Calculation Table, look at the numbers carefully to choose almost the same numbers. You can find two series of the same numbers. $d_1 \times 1 = d_4 \times 2 = d_6 \times 3 = d_{10} \times 4 = d_{16} \times 5 = d_{19} \times 6 = 20.390$; $d_2 \times 1 = d_7 \times 2 = d_{12} \times 3 = d_{18} \times 4 = d_{20} \times 5 = 18.558$. Press the [AVERAGE] button to

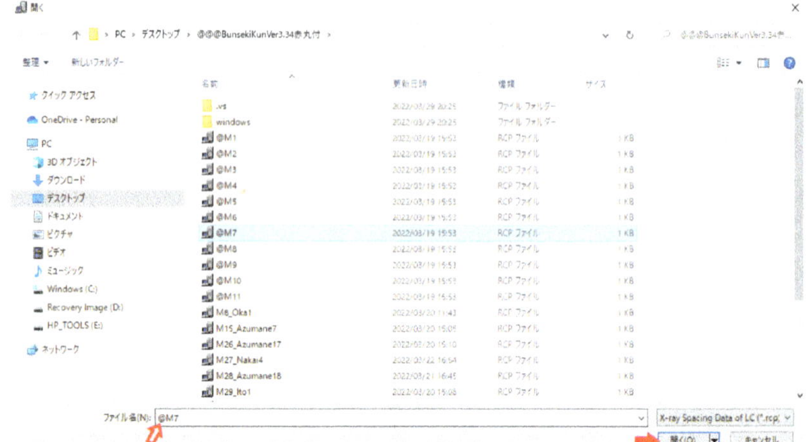

calculate the average values. When you use this subprogram, choose the column name (A, B, C. . .) and the line number ((1), (2), (3). . ..).

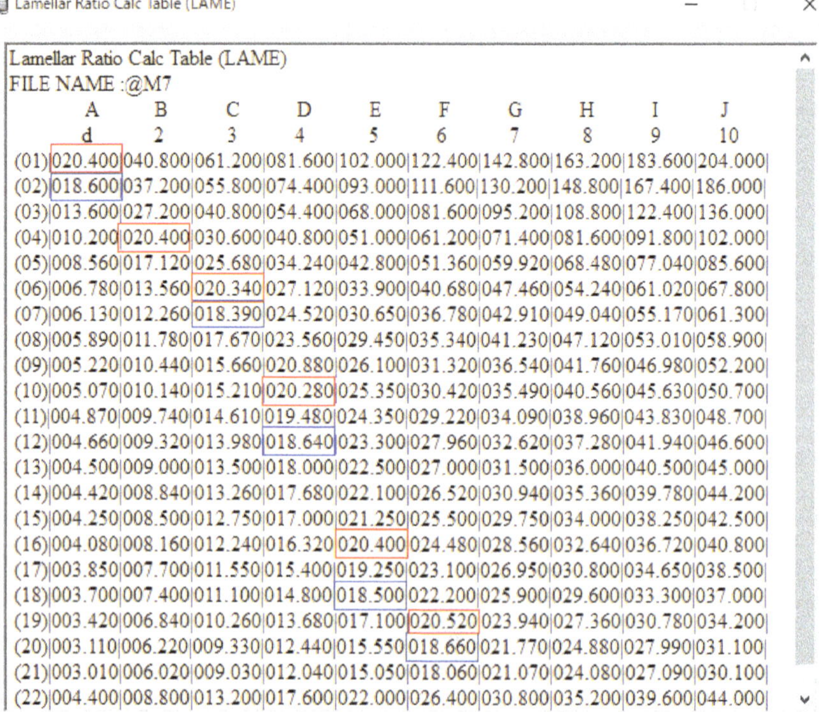

Lamellar Ratio Calc Table (LAME)

FILE NAME :@M7

	A	B	C	D	E	F	G	H	I	J
	d	2	3	4	5	6	7	8	9	10
(01)	020.400	040.800	061.200	081.600	102.000	122.400	142.800	163.200	183.600	204.000
(02)	018.600	037.200	055.800	074.400	093.000	111.600	130.200	148.800	167.400	186.000
(03)	013.600	027.200	040.800	054.400	068.000	081.600	095.200	108.800	122.400	136.000
(04)	010.200	020.400	030.600	040.800	051.000	061.200	071.400	081.600	091.800	102.000
(05)	008.560	017.120	025.680	034.240	042.800	051.360	059.920	068.480	077.040	085.600
(06)	006.780	013.560	020.340	027.120	033.900	040.680	047.460	054.240	061.020	067.800
(07)	006.130	012.260	018.390	024.520	030.650	036.780	042.910	049.040	055.170	061.300
(08)	005.890	011.780	017.670	023.560	029.450	035.340	041.230	047.120	053.010	058.900
(09)	005.220	010.440	015.660	020.880	026.100	031.320	036.540	041.760	046.980	052.200
(10)	005.070	010.140	015.210	020.280	025.350	030.420	035.490	040.560	045.630	050.700
(11)	004.870	009.740	014.610	019.480	024.350	029.220	034.090	038.960	043.830	048.700
(12)	004.660	009.320	013.980	018.640	023.300	027.960	032.620	037.280	041.940	046.600
(13)	004.500	009.000	013.500	018.000	022.500	027.000	031.500	036.000	040.500	045.000
(14)	004.420	008.840	013.260	017.680	022.100	026.520	030.940	035.360	039.780	044.200
(15)	004.250	008.500	012.750	017.000	021.250	025.500	029.750	034.000	038.250	042.500
(16)	004.080	008.160	012.240	016.320	020.400	024.480	028.560	032.640	036.720	040.800
(17)	003.850	007.700	011.550	015.400	019.250	023.100	026.950	030.800	034.650	038.500
(18)	003.700	007.400	011.100	014.800	018.500	022.200	025.900	029.600	033.300	037.000
(19)	003.420	006.840	010.260	013.680	017.100	020.520	023.940	027.360	030.780	034.200
(20)	003.110	006.220	009.330	012.440	015.550	018.660	021.770	024.880	027.990	031.100
(21)	003.010	006.020	009.030	012.040	015.050	018.060	021.070	024.080	027.090	030.100
(22)	004.400	008.800	013.200	017.600	022.000	026.400	030.800	035.200	039.600	044.000

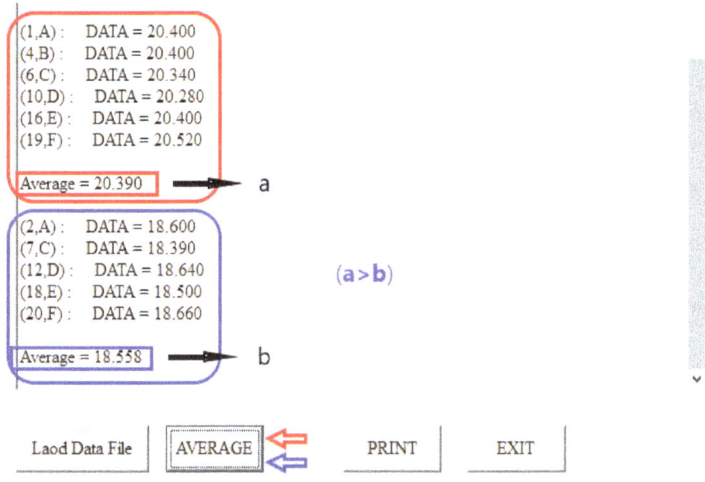

If necessary, press the [PTINT] button.
Finally press the [EXIT] button.

Step (6) When you return to the initial screen [Application Menu] of the Bunseki-kun Ver.3, press sequentially the buttons, [Recip] → [Start] → [Read File] → select the X-ray data of @M7 → [Open(O)].

Then press sequentially [Finish] → [No(N)]. Appeared [Do you calculate the lattice constants?], press the [No(N) button.

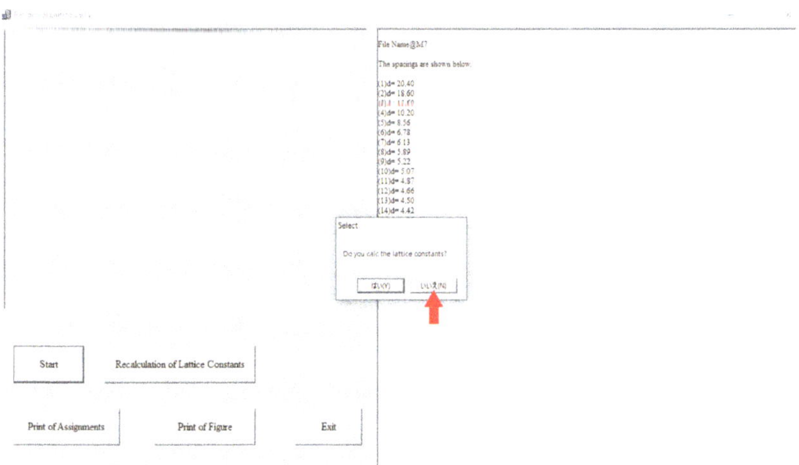

Step (7) When appeared this screen, input the a and b lattice constant values obtained in Step (5). Then, input 90 as the g value and finally the [NEXT] button.

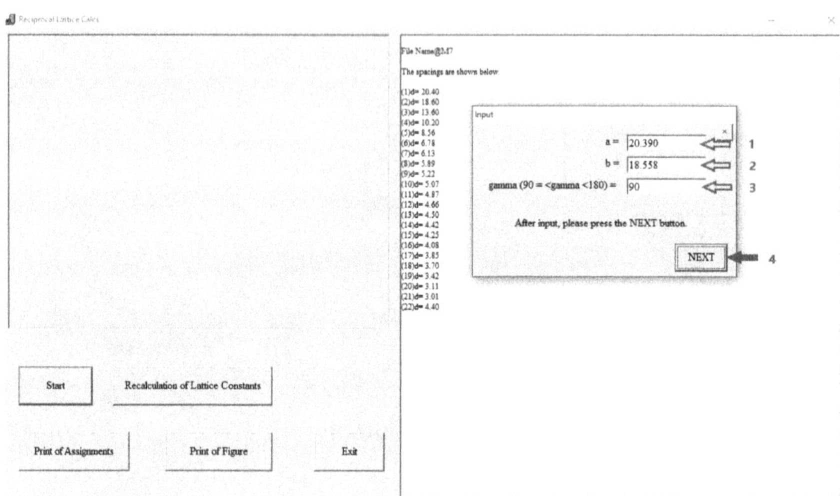

Step (8) When this screen appears, input the numbers of division range, h and k, and then press the [NEXT] button.

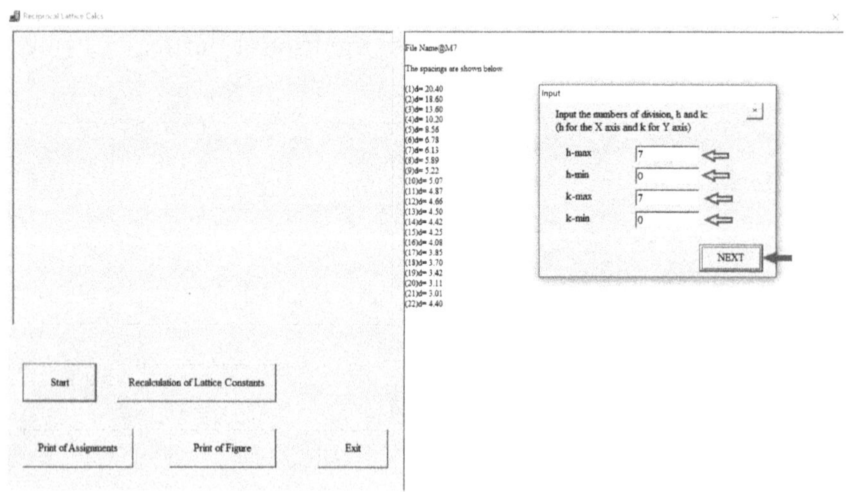

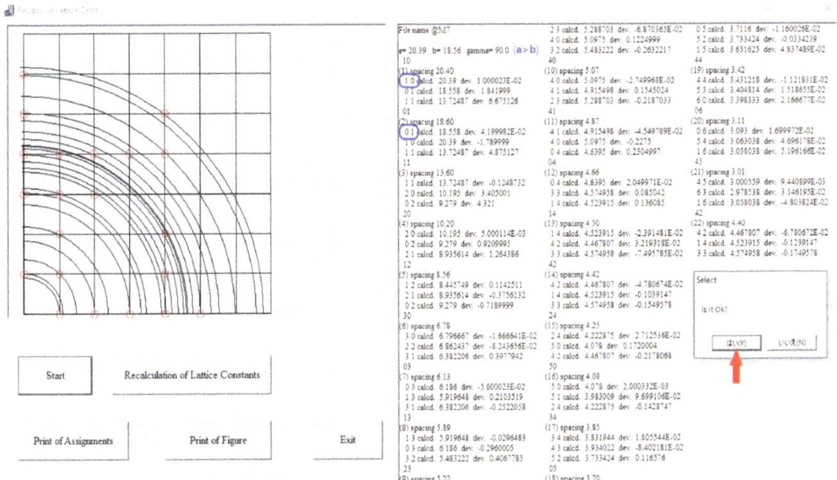

As can be seen from this figure and table, each of the quarter circles fits very well on the intersection points of the reciprocal lattice.

Step (9) If you are satisfied with this lattice calculation, press the [Yes(Y)] button.

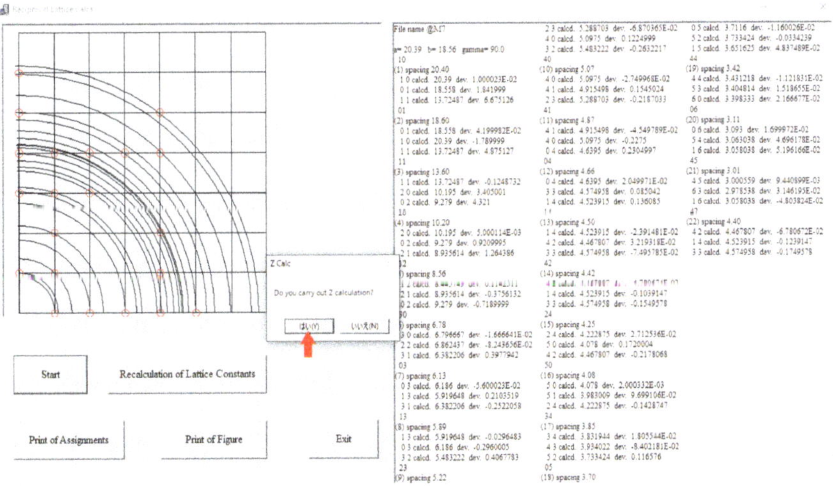

Step (10) When appeared the question of "Do you carry out Z value calculation?," press the [Yes(Y)] button.

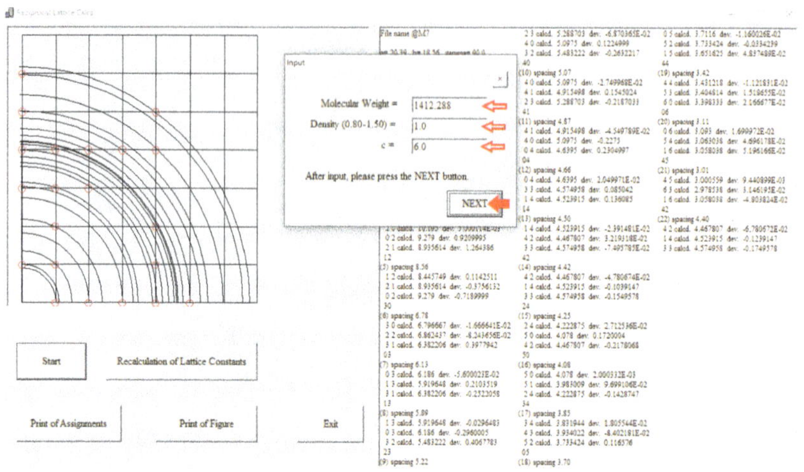

Step (11) Input the molecular weight, an assumed density and the c value in the inset table.

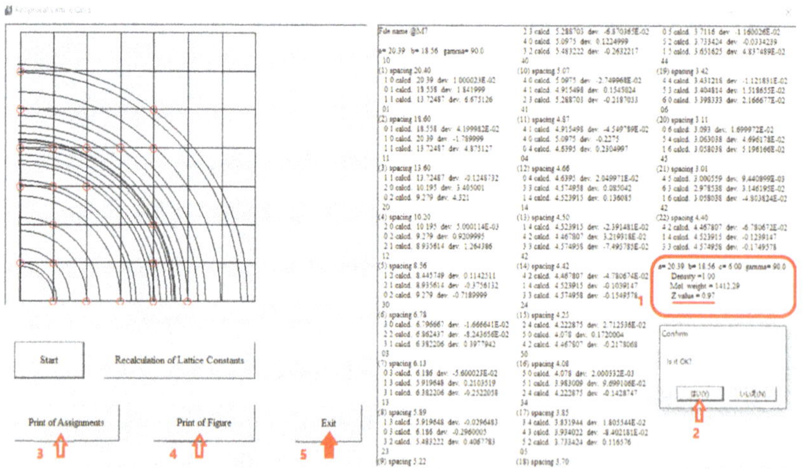

Thus, you obtain Z = 0.97 ≅1. The Z value is integer 1, so that this identification of $\mathrm{Col_{rd}}$ is consistently correct.

Four symmetries are known so far for two-dimensional rectangular lattices, as shown in Figure 3. Among them, the symmetries with Z = 1 is P2m.

From Figure 3, there is no extinction rule for this P2m symmetry. Accordingly, this liquid crystalline phase gives many reflection lines in the X-ray diffraction pattern in comparison with the other symmetries of rectangular lattice.

If you are OK for the confirmation, press the [Yes(Y)] button. If necessary, press the [Print of Figure] and [Print of Assignments] buttons. Finally, press the [EXIT] button. The liquid crystal structure analysis results are summarized in Table 8.

Table 8. X-ray data of @M7 and the indexation of the reflections.

Entry No.	M7		
Reference	p56, $(2\text{-Et-}C_6)_8PcH_2$		
Mw	Mw=1414.288		
Temperature	55°C		
Mesophase	Col_{rd}(P2m)		
Lattice constants	a=70.8, b=32.6		
Peak No.	$d_{obs.}$	$d_{calcd.}$	*(hkl)*
1	20.4	20.4	(1 0 0)
2	18.6	18.6	(0 1 0)
3	13.6	13.7	(1 1 0)
4	10.2	10.2	(2 0 0)
5	8.56	8.45	(1 2 0)
6	6.78	6.80	(3 0 0)
7	6.13	6.19	(0 3 0)
8	5.89	5.92	(1 3 0)
9	5.22	5.29	(2 3 0)
10	5.07	5.10	(4 0 0)
11	4.87	4.92	(4 1 0)
12	4.66	4.64	(0 4 0)
13	4.50	4.52	(1 4 0)
14	4.42	4.47	(4 2 0)
15	4.25	4.22	(2 4 0)
16	4.08	4.08	(5 0 0)
17	3.85	3.84	(3 4 0)
18	3.70	3.71	(0 5 0)
19	3.42	3.43	(4 4 0)
20	3.11	3.09	(0 6 0)
21	3.01	3.00	(4 5 0)
22	*ca.* 4.4	–	#

7.1 How to analyse @(20)Azumane11 $Col_{tet.o}$

7.1.1 *Loading X-ray diffraction data*

Step (1) In the initial screen [Application Menu] of the Bunseki-kun Ver.3, press sequentially the buttons, [Recip] → [Start] → [Read File], in order to download the X-ray data of @(20)Azumane11.

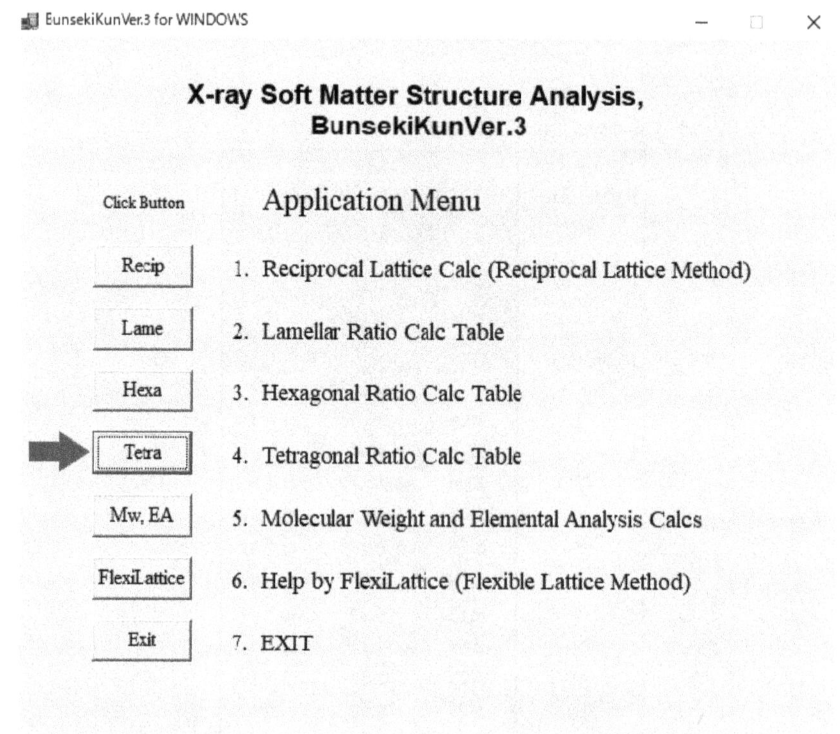

Peak No.	$d_{obs.}$
1	29.7
2	21.4
3	9.39
4	8.30
5	7.27
6	6.78
7	4.96
8	ca.4.8
9	4.75
10	3.82

$$\mathbf{d_1 : d_2 = 1 : 1/\sqrt{2}}$$

Step (2) Look at the X-ray data and estimate what kind of dimensionality the liquid crystal has. Judging from Article 1 to Article 4 in the golden rules of liquid crystal structure analysis like as the previous examples, this phase may have a tetragonal lattice, because the first two spacings show $d_1 : d_2 = 1 : 1/\sqrt{2}$ in accordance with Article 3.

Step (3) Therefore, press sequentially the buttons, [X] in the right up → [EXIT], to return to the initial screen [Application Menu]. Press the [Tetra] button. After pressed the button, a following tetragonal spreadsheet appears.

Tetragonal Ratio Calc Table (TETRA)

Tetragonal Ratio Calc Table (TETRA)

	d	SQR02	SQR04	SQR05	SQR08	SQR09	SQR10	SQR13	SQR16	SQR17
(01)	029.700	042.002	059.400	066.411	084.004	089.100	093.920	107.085	118.800	122.456
(02)	021.400	030.264	042.800	047.852	060.528	064.200	067.673	077.159	085.600	088.234
(03)	009.390	013.279	018.780	020.997	026.559	028.170	029.694	033.856	037.560	038.716
(04)	008.300	011.738	016.600	018.559	023.476	024.900	026.247	029.926	033.200	034.222
(05)	007.270	010.281	014.540	016.256	020.563	021.810	022.990	026.212	029.080	029.975
(06)	006.780	009.588	013.560	015.161	019.177	020.340	021.440	024.446	027.120	027.955
(07)	004.960	007.014	009.920	011.091	014.029	014.880	015.685	017.884	019.840	020.451
(08)	004.800	006.788	009.600	010.733	013.576	014.400	015.179	017.307	019.200	019.791
(09)	004.750	006.718	009.500	010.621	013.435	014.250	015.021	017.126	019.000	019.585
(10)	003.820	005.402	007.640	008.542	010.805	011.460	012.080	013.773	015.280	015.750

	d	SQR18	SQR20	SQR25	SQR26	SQR29	SQR32	SQR34	SQR36	SQR37
(01)	029.700	126.006	132.822	148.500	151.441	159.939	168.009	173.179	178.200	180.658
(02)	021.400	090.793	095.704	107.000	109.119	115.243	121.057	124.782	128.400	130.171
(03)	009.390	039.838	041.993	046.950	047.880	050.567	053.118	054.753	056.340	057.117
(04)	008.300	035.214	037.119	041.500	042.322	044.697	046.952	048.397	049.800	050.487
(05)	007.270	030.844	032.512	036.350	037.070	039.150	041.125	042.391	043.620	044.222
(06)	006.780	028.765	030.321	033.900	034.571	036.511	038.353	039.534	040.680	041.241
(07)	004.960	021.043	022.182	024.800	025.291	026.710	028.058	028.922	029.760	030.171
(08)	004.800	020.365	021.466	024.000	024.475	025.849	027.153	027.989	028.800	029.197
(09)	004.750	020.153	021.243	023.750	024.220	025.580	026.870	027.697	028.500	028.893
(10)	003.820	016.207	017.084	019.100	019.478	020.571	021.609	022.274	022.920	023.236

Load Data File   PRINT   EXIT

As can be seen from this table, the circled values are almost constant aligned with the average value of 29.949 Å.

Therefore, the ratios of spacing values clearly show that the liquid crystal phase has a two-dimensional tetragonal structure, from Article 3 in the "golden rules for liquid crystal structure analysis."

Hereupon, if necessary, press the [PRINT] button. Then, press the [EXIT] button.

Step (4) Loading X-ray diffraction data. When pressed the [EXIT] button, you return to the initial screen [Application Menu]. In order to download again the X-ray data of @(20)Azumane11, press sequentially the buttons, [Recip] → [Start] → [Read File] → [FINISH] → [No(N)]. After pressed the [No(N)] button, you will be asked as "Do you calculate the lattice constants?". Press [No(N)]. A screen appears as follows.

Step (5) When the inset screen appeared, input the tetragonal lattice constant a = b = 29.949 Å, which was obtained as the average value between Step (3) and Step (4). Then input $\gamma = 90°$.

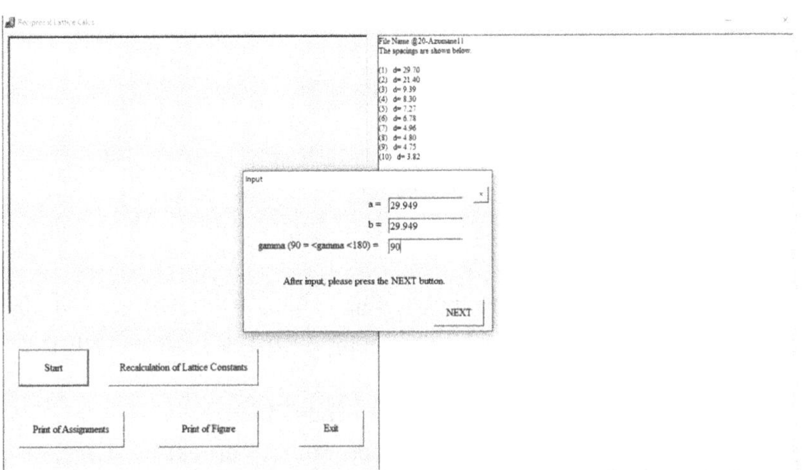

Step (6) Input the number of divisions, h and k. When you press the [NEXT] button, the next screen appears.

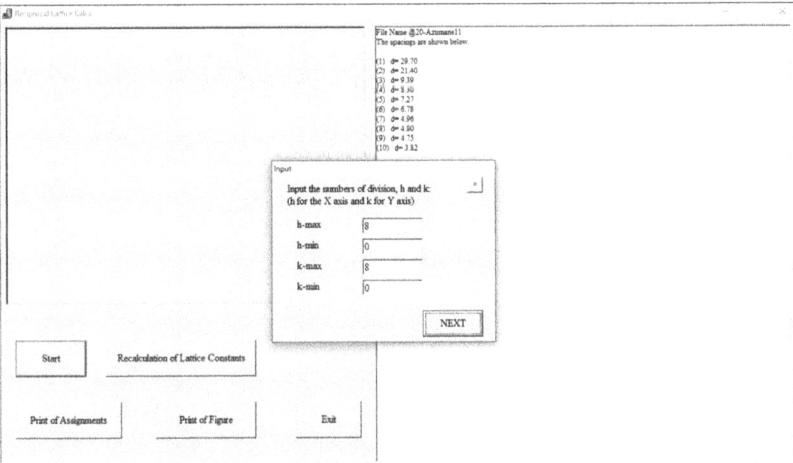

A reciprocal lattice plane will be drawn in the next screen.

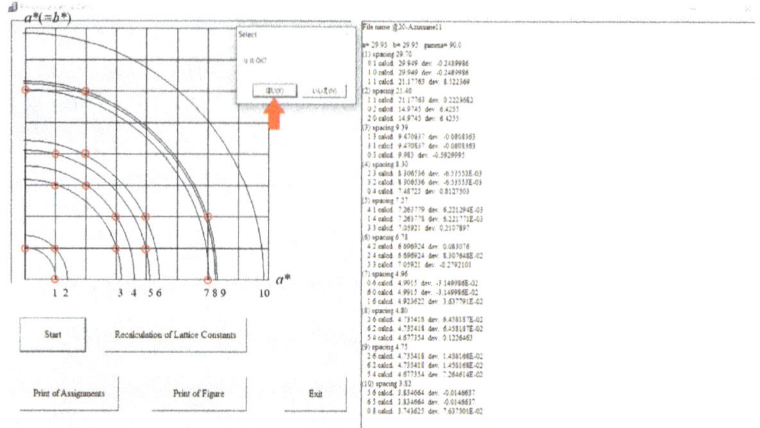

As can be seen from the figure and table, the reflections of Peak Nos. 1 to 7 and 9 are in good agreement with the intersection points marked with small red circles.

Peak No. 8 corresponds to the average distance of molten alkyl groups and Peak No. 10 corresponds to the stacking distance of discotic molecules in the column. Therefore, this liquid crystal phase can be identified as a tetragonal ordered columnar ($Col_{tet.o}$) phase.

Step (7) If it is OK for you, press the [Yes(Y)] button.

Step (8) Input the molecular weight, an assumed density and the c value in the inset table.

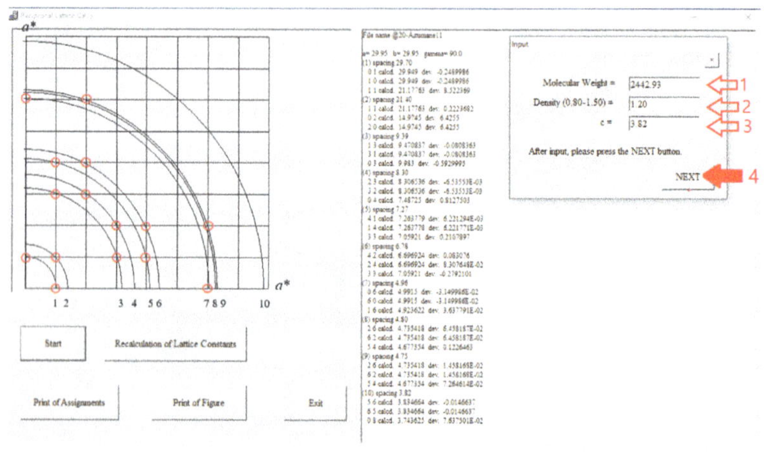

Verification by Z value calculation

If Z is the number of molecules in the unit cell of the $\mathrm{Col}_{\mathrm{tet.o}}$ phase, Z = 1 in the $\mathrm{Col}_{\mathrm{tet.o}}$ phase. Therefore, if the identification of this liquid crystal phase is correct, the Z value obtained above and calculated from the lattice constant must be an integer value of 1. Whether or not this identification is correct is verified by the following Z value calculation.

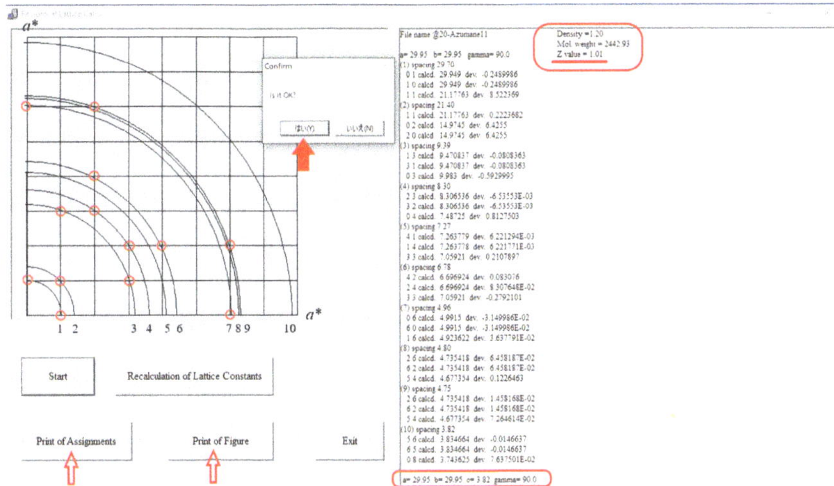

As can be seen the result in the right table, the Z value was obtained to be an integer of 1. Therefore, you can confirm that the above identification is correct. Thus, this liquid crystal phase can be identified as a [2D⊕1D]-dimensional tetragonal ordered columnar phase ($\mathrm{Col}_{\mathrm{tet.o}}$).

The analysis results are summarized and tabulated as follows.

Table 9. X-ray data of (20)Azumane11 and the indexation of the reflections.

Entry No.	◎ (20)Azumane11		
Reference	p74, 1b, $(C_{10}OPh)_8PzCu$ =		
Mw	Mw=2442.93		
Temperature	140°C		
Mesophase	$Col_{tet.o}$		
Lattice constants	a = 29.95, h = 3.82; Z = 1.0 for ρ = 1.20		
Peak No.	$d_{obs.}$	$d_{calcd.}$	*(hkl)*
1	29.7	29.9	(1 0 0)
2	21.4	21.2	(1 1 0)
3	9.39	9.47	(1 3 0)
4	8.30	8.31	(2 3 0)
5	7.27	7.26	(4 1 0)
6	6.78	6.70	(2 4 0)
7	4.96	4.99	(6 0 0)
8	ca.4.8	–	#
9	4.75	4.74	(6 2 0)
10	3.82	–	h: (0 0 1)

7.2 How to analyse @(99)Ariyoshi7 $Col_{tet.o}$

7.2.1 *Loading X-ray diffraction data*

Step (1) In the initial screen [Application Menu] of the Bunseki-kun Ver.3, press sequentially the buttons, [Recip] → [Start] → [Read File], in order to download the X-ray data of @(99)Ariyoshi7.

Step (2) Look at the X-ray data and estimate what kind of dimensionality the liquid crystal has. Judging from Article 1 to Article 4 in the golden rules of liquid crystal structure analysis like as the previous examples, this phase may have a tetragonal lattice, because the first two spacings show $d_1 : d_2 = 1 : 1/\sqrt{2}$ in accordance with Article 3.

Table 10. X-ray data of (99)Ariyoshi7.

Peak No.	$d_{obs.}$
1	**32.3**
2	**23.0**
3	**ca.4.7**
4	**ca.3.5**

$\mathbf{d_1 : d_2 = 1 : 1/\sqrt{2}}$

Step (3) Therefore, press sequentially the buttons, [Recip] → [Start] → [Read File] → [FINISH] → [No(N)]. After pressed the [No(N)] button, you will be asked as "Do you calculate the lattice constants?". Press [Yes(Y)] and then [2D-tetragonal]. A screen appears as follows.

Step (4) Choose Peak No. 1 as the standard peak. Then, input (10) as the (hk) value of this peak.

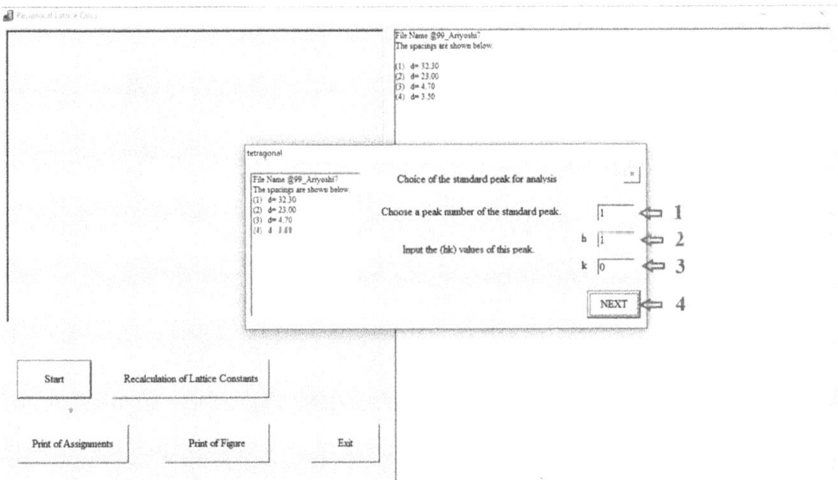

Step (5) Input the numbers of division, h and k.

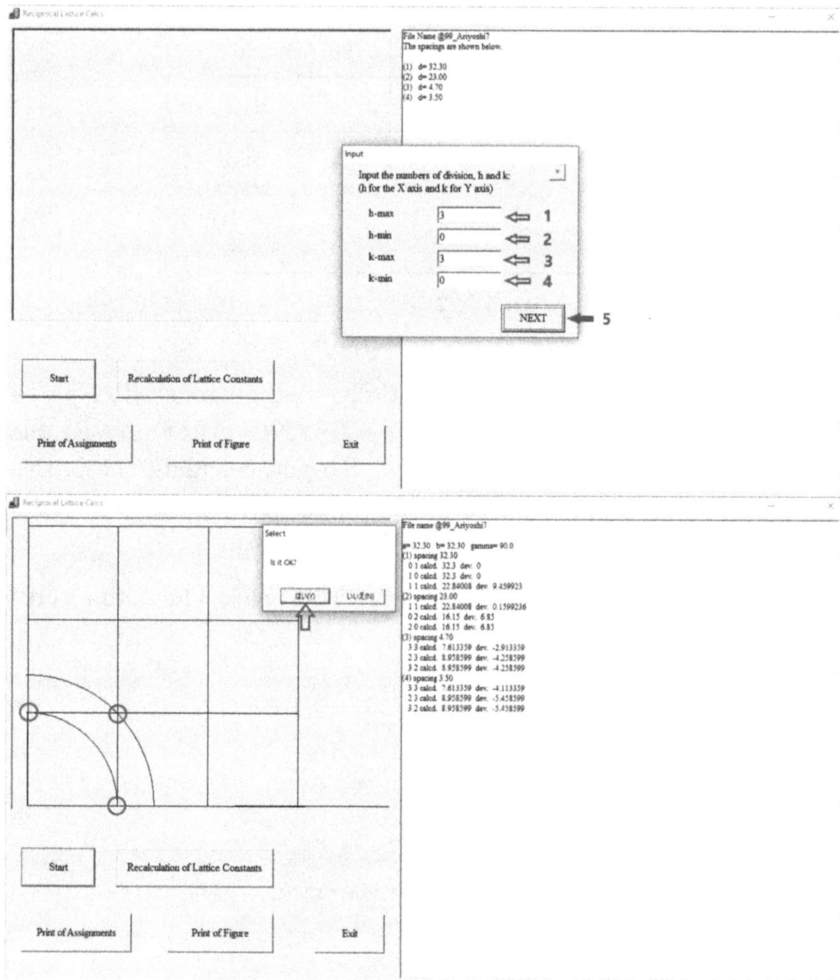

As can be seen from this figure and table, the reflections of peaks Nos. 1 and 2 are in good agreement with the intersection points marked with small red circles.

Therefore, this liquid crystal phase can be identified as a tetragonal ordered columnar ($Col_{tet.o}$) phase.

Step (6) If it is OK for you, press the [Yes(Y)] button.

7.2.2 *Verification by Z value calculation*

If the number of molecules in a unit cell of $Col_{tet.o}$ phase is set as Z, the Z value should be equal to 1. Therefore, if the identification of this liquid crystal phase is correct, the Z value calculated from the lattice constant must be an integer value of 1. If it is not an integer, such as 0.5 or 1.5, the present identification is wrong. Therefore, it can be verified whether this identification is correct or not, by using the Z-value calculation.

Step (7) When you press the [Yes(Y)] button in the above (6) step, a following screen appears for verification by Z value calculation. For the question [Do you carry out Z calculation?], click the [Yes(Y)] button.

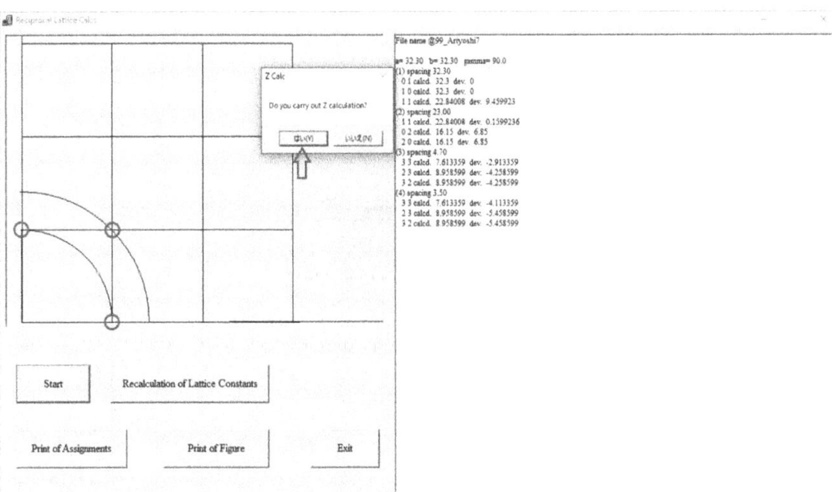

Step (8) Input the molecular weight, an assumed density and the c value in the inset table.

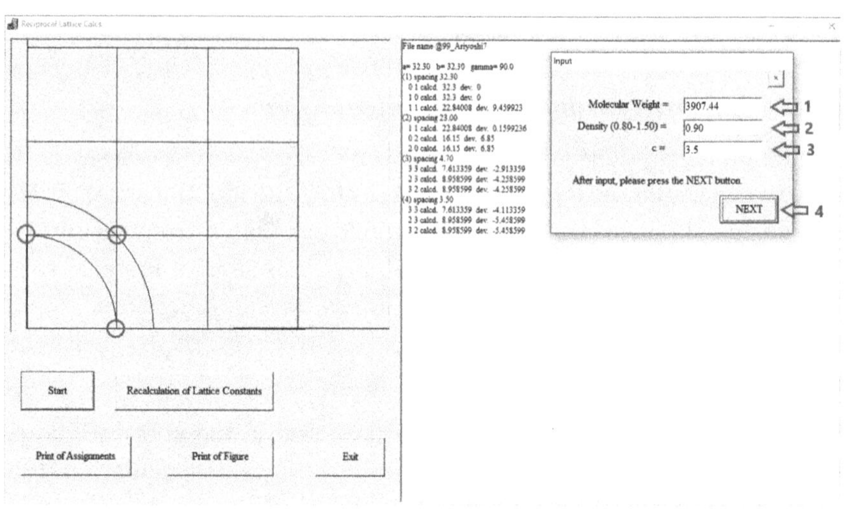

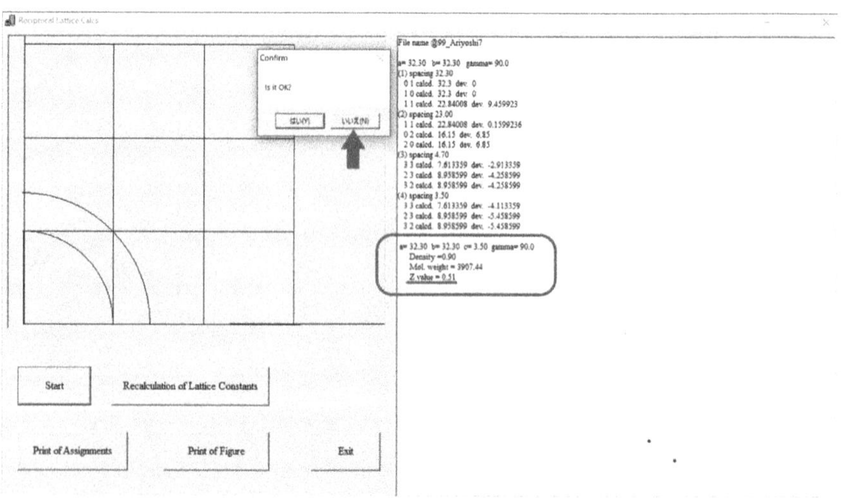

As can be seen the result in this table, the obtained Z value is not an integer of 1 but 0.51. Therefore, the above identification is wrong.

Step (9) Therefore, press the [No(N)] button.

Step (10) Choose Peak No. 1 as the standard peak. Then, input (**11**) as the (hk) value of this peak.

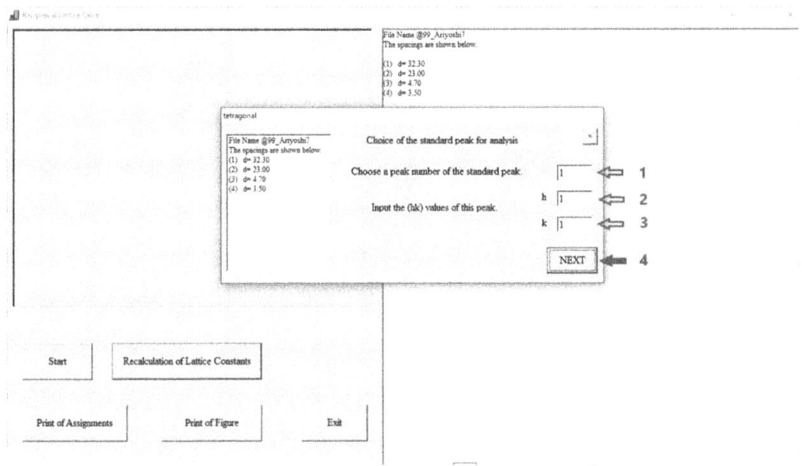

Step (11) Input the numbers of division, h and k.

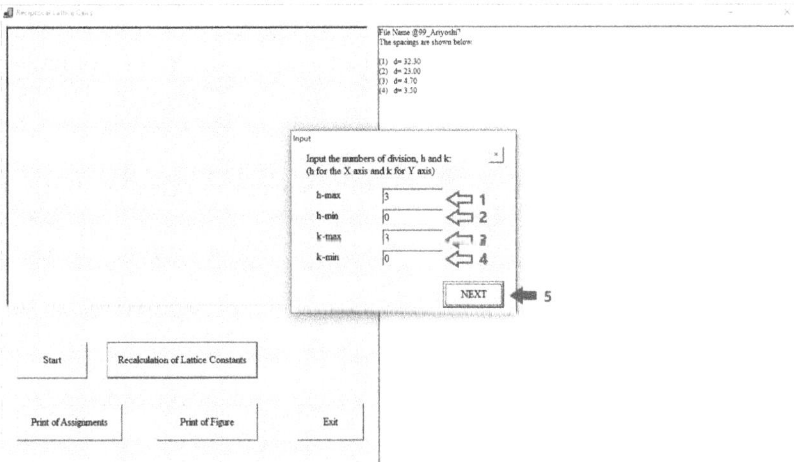

This wrong analysis is attributable to the present assumption that Peak No. 1 was used as a standard peak reflected from the (**10**) plane. Considering deeply on Article 3 of "Golden Measure of Liquid Crystal Structure Analysis", $d_1 : d_2 = 1 : 1/\sqrt{2}$ is equivalent to

$d_1 : d_2 = 1/\sqrt{2} : 1/2$. Therefore, you try again the analysis by assuming that Peak No. 1 is as a standard peak reflected from the (**11**) plane.

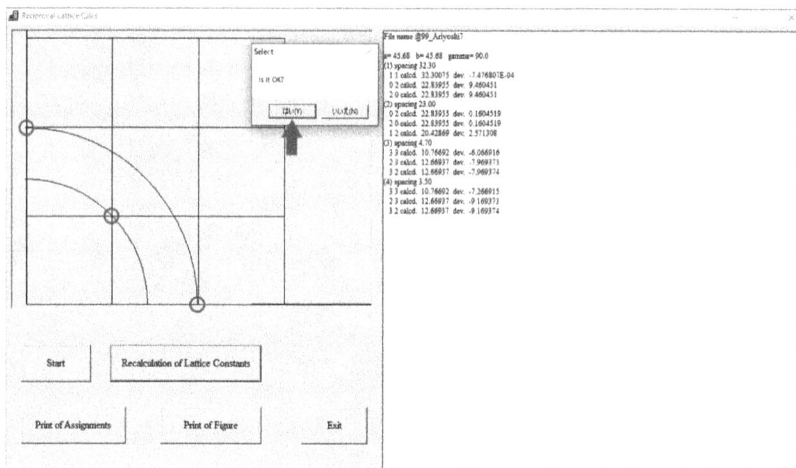

As can be seen from this figure and table, the reflections of peaks Nos. 1 and 2 are in good agreement with the intersection points marked with small red circles.

Therefore, the second assumption also gives the same conclusion that this liquid crystal phase can be identified as a tetragonal ordered columnar ($Col_{tet.o}$) phase.

Step (12) If you are OK for this result, press the [Yes(Y)] button.

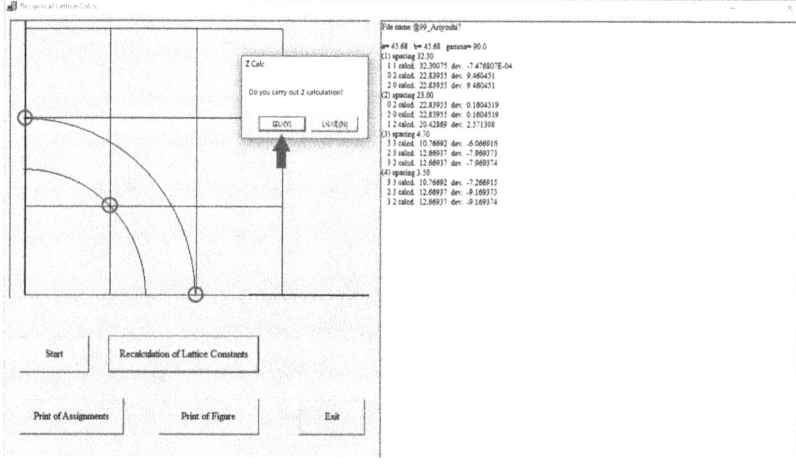

Step (13) When you press the [Yes(Y)] button in the above (12) step, a above screen appears for verification by Z value calculation. For the question [Do you carry out Z calculation?], click the [Yes(Y)] button.

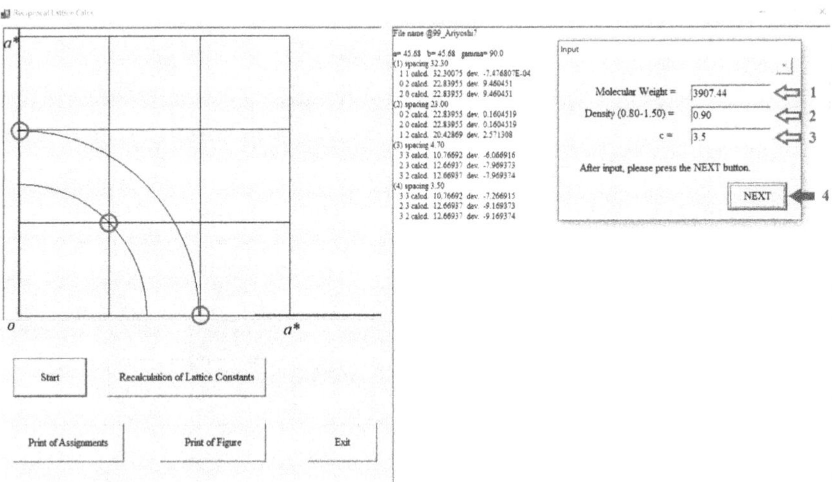

Step (14) Input the molecular weight, an assumed density and the c value in the inset table.

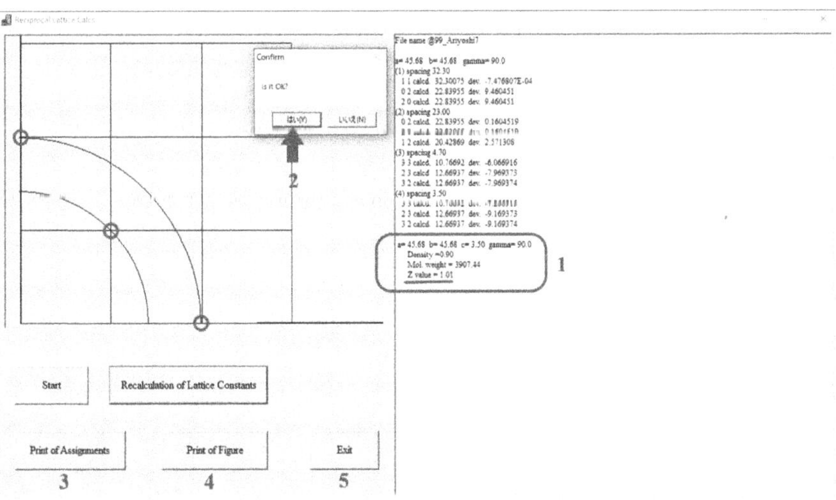

Accordingly, you can obtain as Z = 1.01 ≅ 1.0, as shown here. Thus, the calculated Z value is integer 1, so that this identification of $\mathrm{Col}_{\mathrm{tet.o}}$ is consistently correct.

Note: As can be seen from this example, verification by Z value calculation is essential at the final step of liquid crystal structure analysis.

Step (15) If you are OK for the confirmation, press the [Yes(Y)] button. If necessary, press the [Print of Assignments] and [Print of Figure] buttons. Finally, press the [EXIT] button.

The analysis results are summarized and tabulated as follows.

Table 11. X-ray data of (99)Ariyoshi7 and the indexation of the reflections.

Entry No.	◎ (99)Ariyoshi7		
Reference	p146, $C_{14}(OC_{12}OH)PcCu$: 5c		
Mw	Mw=3907.44		
Temperature	155.0°C		
Mesophase	$\mathrm{Col}_{\mathrm{tet\cdot o}}$		
Lattice constants	a = 45.7, h = ca. 3.5, Z = 1.0 for ρ = 0.90		
Peak No.	$d_{obs.}$	$d_{calcd.}$	(*hkl*)
1	32.3	32.3	(1 1 0)
2	23.0	22.8	(2 0 0)
3	ca.4.7	–	#
4	ca.3.5	–	h: (0 0 1)

Section 2. Structure Analysis of Smectic Liquid Crystal Phases

8 How to Analyse @M9_S$_A$

8_How to analyse @M9_S$_A$
8-1. Loading X-ray diffraction data

Step (1) In the initial screen [Application Menu] of the Bunseki-kun Ver.3, press sequentially the buttons, [Recip] → [Start] → [Read File], in order to download the X-ray data of @M9.

Step (2) Look at the X-ray data. You can see that it has a series of ratios specific to simple lamellar structure, as follows:

$$\mathrm{d}_1 : \mathrm{d}_2 : \mathrm{d}_3 = 32.8 : 16.7 : 11.2 = 1 : \frac{1}{2} : \frac{1}{3}$$

Peak No. 4 is very broad and attributable to the molten alkyl chains, so that this peak is out of our consideration from the golden rules of liquid crystal structure analysis to estimate the dimensionality. Thus, since no other reflection is observed, this phase can be estimated to be smectic A phase or C phase having only a 1D lamellar structure.

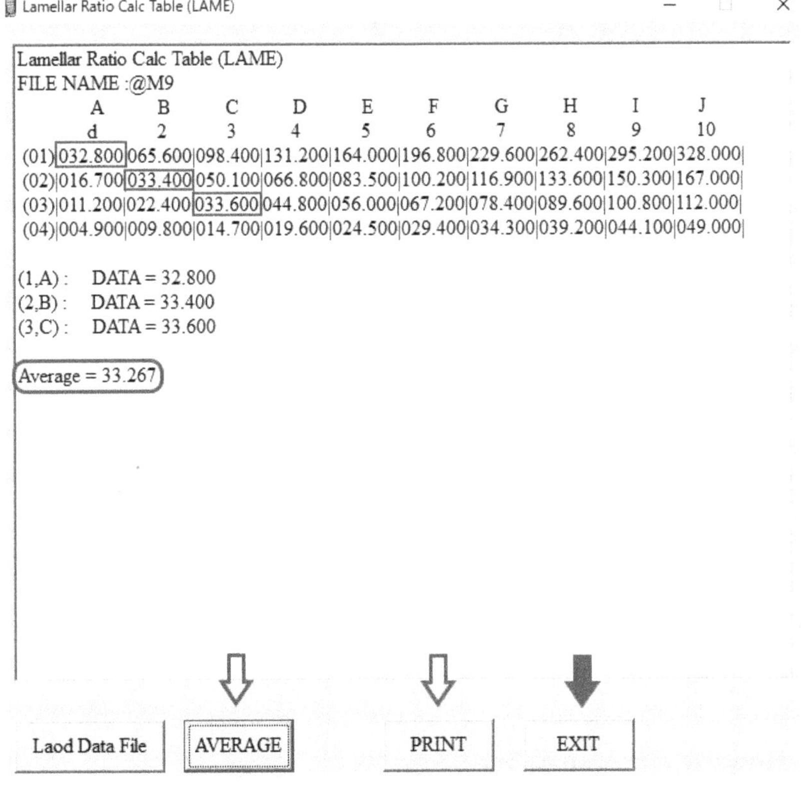

Step (3) To obtain the lamellar thickness, you use the subprogram [Lame] in the initial screen [Application Menu]. Press the [Lame] button and then click the [Load Data File] button. Select the target file @M9 to download. A right up the lamellar ratio calculation table appears.

However, it cannot be determined any more from X-ray structure analysis which phase it has, so that the natural texture of this liquid crystal phase should be observed with a polarization microscope. Since it exhibits a fan-shaped texture and a pseudo-isotropic region characteristic to smectic A phase, this liquid crystal phase is finally identified as a smectic A phase.

Note: Thus, X-ray structural analysis is not almighty for the identification of liquid crystal phases. Therefore, it is necessary to also observe the texture with a polarizing microscope at the same time, in order to make an overall judgment.

The liquid crystal structure analysis results are summarized and tabulated as follows.

Table 12. X-ray data of @M9 and the indexation of the reflections.

Entry No.	M9		
Reference	p45, $(C_{12}Salen)_2Ni$		
Mw	Mw=633.57		
Temperature	200°C		
Mesophase	S_A		
Lattice constants	c = 33.3		
Peak No.	$d_{obs.}$	$d_{calcd.}$	*(hkl)*
1	32.8	32.8	(0 0 1)
2	16.7	16.6	(0 0 2)
3	11.2	11.1	(0 0 3)
4	*ca.* 4.9	–	#

#: halo of molten alkyl chains

9 How to Analyse @M10_S$_E$

9.1 Estimation of dimensionality of the mesophase

Step (1) In the initial screen [Application Menu] of the Bunseki-kun Ver.3, press sequentially the buttons, [Recip] → [Start] → [Read File], in order to download the X-ray data of @M10.

When you look at the X-ray data. you can see that it has a series of ratios specific to simple lamellar structure, as follows:

$$\mathrm{d}_1 : \mathrm{d}_2 = 25.2 : 13.1 = 1 : \frac{1}{2}$$

Step (2) In order to certify the lamellar ratio, you should return to the initial screen, Application Menu, press the [Lame] button, and then click the [Load Data File] button. When you select the target file, @M10, a following screen appears.

As can be seen the simple ratio from this screen, you can certify that this phase is a liquid crystal phase having a lamellar structure having an interlayer distance c = 25.7 Å.

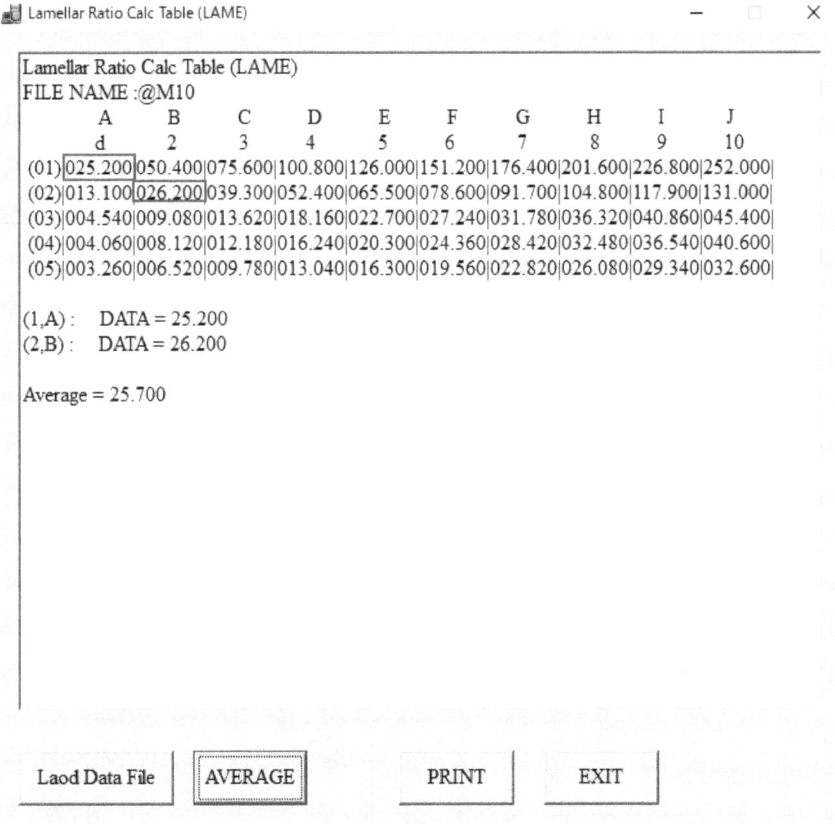

Lamellar Ratio Calc Table (LAME)

Lamellar Ratio Calc Table (LAME)
FILE NAME :@M10

	A	B	C	D	E	F	G	H	I	J
	d	2	3	4	5	6	7	8	9	10
(01)	025.200	050.400	075.600	100.800	126.000	151.200	176.400	201.600	226.800	252.000
(02)	013.100	026.200	039.300	052.400	065.500	078.600	091.700	104.800	117.900	131.000
(03)	004.540	009.080	013.620	018.160	022.700	027.240	031.780	036.320	040.860	045.400
(04)	004.060	008.120	012.180	016.240	020.300	024.360	028.420	032.480	036.540	040.600
(05)	003.260	006.520	009.780	013.040	016.300	019.560	022.820	026.080	029.340	032.600

(1,A) : DATA = 25.200
(2,B) : DATA = 26.200

Average = 25.700

Laod Data File | AVERAGE | PRINT | EXIT

However, this liquid crystal phase shows additional three sharp peaks at 4.54 Å, 4.06 Å, and 3.26 Å in the higher angle region. Therefore, this phase is neither a smectic A phase nor C phase having a simple lamellar structure seen in the above example (See @M9_ S_A in Section 8). Although the present phase (@M10) shows the 1D reflections in the low angle region, three additional 2D reflections appear in the high angle region.

Therefore, you should consider the spacings in the high angle region to estimate the dimensionality of this phase. When the X-ray data in the high angle region are checked sequentially from the first article of the "**Golden rule for liquid crystal structure analysis**" similarly to the previous columnar liquid crystals. Accordingly, the following calculations are carried out.

$$\text{1D-lamellar } 4.56\ \text{Å} \div 2 = 2.28\ \text{Å} \tag{1}$$

$$\text{2D-hexagonal } 4.56\ \text{Å} \div \sqrt{3} = 2.63\ \text{Å} \tag{2}$$

$$\text{2D-tetragonal } 4.56\ \text{Å} \div \sqrt{2} = 3.22\ \text{Å} \tag{3}$$

You examine these calculated values with the observed values sequentially from Articles 1 ~ 4 and Equations (1) ~ (3) as follows:

(i) From Article 1 and Equation (1), the calculated value of 2.28 Å does not match the observed value.
(ii) Next, considering from Article 2 and Equation (2), the calculated value of 2.63 Å does not match the observed value. Therefore, it does not have a 2D-hexagonal lattice.
(iii) Furthermore, considering Article 3 and Equation (3), the calculated value of 3.22 Å agrees with the observed value of 3.26 Å within experimental error, whereas the observed value of 4.06 Å cannot be explained. Therefore, it does not have a 2D-tetragonal lattice.

Thus, it is neither 1D-lamellar, nor 2D-hexagonal nor 2D-tetragonal. It can be considered to have a 2D-rectangular lattice or a 2D-oblique lattice. Since none of the smectic phases have a 2D-oblique lattice, it is presumed that this liquid crystal phase is a smectic E phase having a 2D-rectangular lattice (cf. Table 3 in this book).

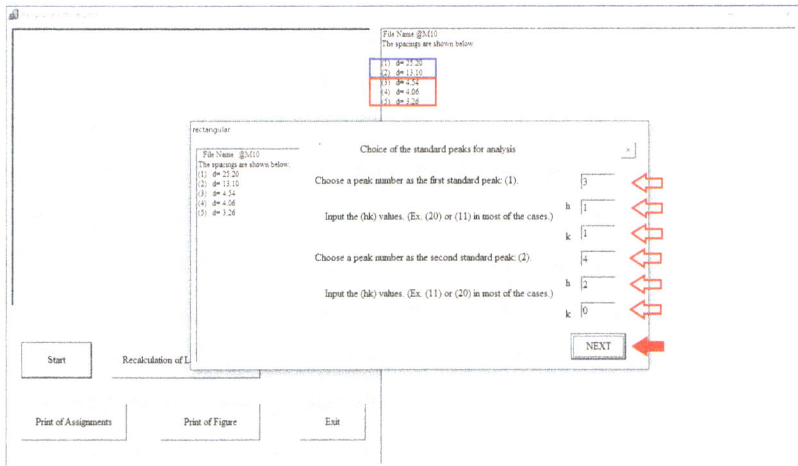

It is usually expected that the strongest reflection intensity from the two-dimensional rectangular lattice is from the (20) and (11) planes, as can be seen from Figure 1.

Step (3) So, choose Peak Nos. 3 and 4 as the two standard peaks, and input (11) and (20) into the (hk) values in the right blanks of the inset screen.

Then, press [NEXT] button.

Step (4) When the next screen, input appropriate numbers of division, h and k. In this case, h: 3, 0 and k: 3, 0.

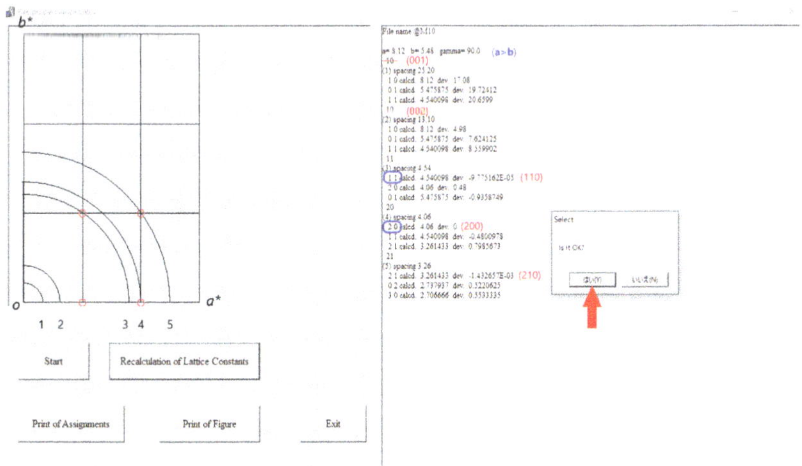

1. Thus, assuming $d_{11} = 4.54\,\text{Å}$ and $d_{20} = 4.06\,\text{Å}$ for becoming $a > b$, the lattice constants are obtained $a = 8.12$ Å and $b = 5.48$ in the right table. A corresponding reciprocal lattice plane and the quarter circles appear in the left figure.
2. As can be seen from the figure and table, the reflections of Peak Nos. 3 to 5 are in good agreement with the intersection points marked with small red circles. Peak Nos. 3 to 5 can be indexed as (11), (20) an (21), respectively.
3. From the extinction rules in Figure 3, it can be concluded that this rectangular lattice has the symmetry of $P2_1/a$. Therefore, this liquid crystal phase can be identified as a smectic E phase with $P2_1/a$ symmetry.

Step (5) If it is OK for you, press the [Yes(Y)] button.

9.2 Verification by Z value calculation

When Z is the number of molecules in the unit cell, Z = 2 in the $S_E(P2_1/a)$ phase. Therefore, if the identification of this liquid crystal phase is correct, the Z value theoretically calculated from these lattice constants must be an integer value 2. Hence, from the following theoretical Z value calculation, you should verify whether this Z value is correct or not.

Step (6) Input the molecular weight, an assumed density and the c value in the inset table.

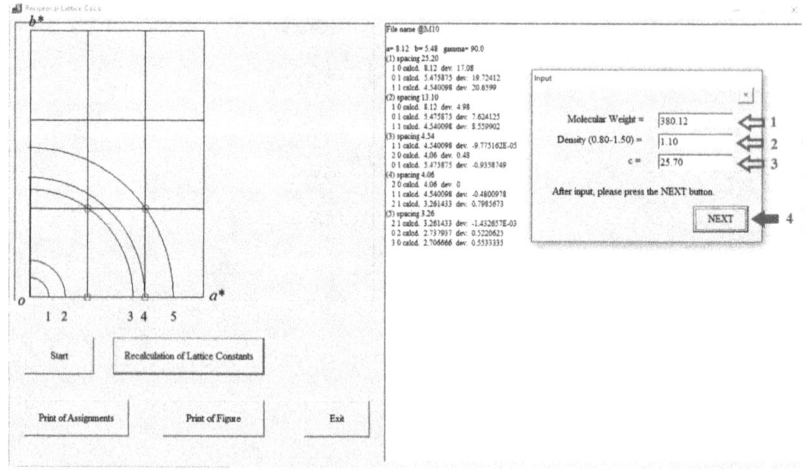

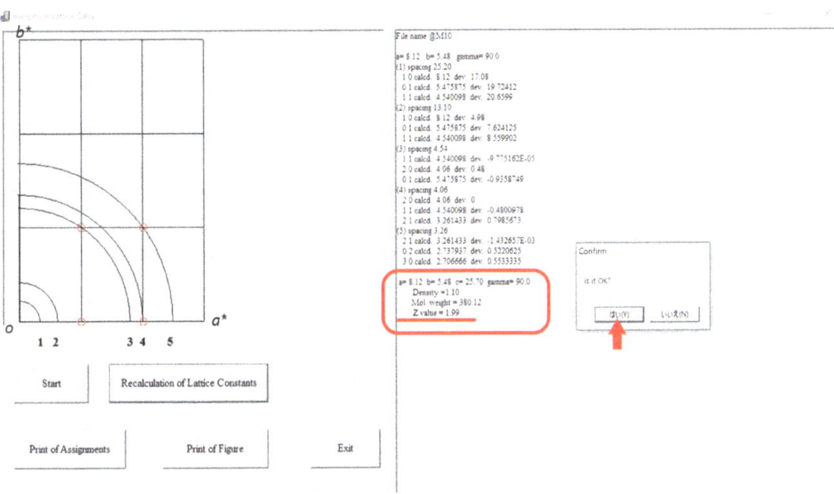

As can be seen the result in the right table, the Z value was obtained to be an integer of 2. Therefore, you can confirm that the above identification is correct. Thus, this liquid crystal phase can be identified as a [2D⊕1D]-dimensional smectic E phase with a $P2_1/a$ symmetry ($S_E(P2_1/a)$).

The analysis results are summarized and tabulated as follows.

Table 13. X-ray data of **@M10** and the indexation of the reflections.

Entry No.	M10		
Reference	p54, 10(n=80)		
Mw	Mw=380.53		
Temperature	122°C		
Mesophase	$S_E(P2_1/a)$		
Lattice constants	a = 8.12, b = 5.48, c = 25.7 Z = 2.0 for ρ = 1.1		
Peak No.	$d_{obs.}$	$d_{calcd.}$	(*hkl*)
1	25.2	25.7	(0 0 1)
2	13.1	12.9	(0 0 2)
3	4.54	4.54	(1 1 0)
4	4.06	4.06	(2 0 0)
5	3.26	3.26	(2 1 0)

10 How to Analyse @M11_S_T

Step (1) In the initial screen [Application Menu] of the Bunseki-kun Ver.3, press sequentially the buttons, [Recip] → [Start] → [Read File], in order to download the X-ray data of @M11.

Step (2) Look at the X-ray data. You can see that it has a series of ratios specific to simple lamellar structure, as follows:

$$32.9 : 16.1 : 10.7 : 7.97 : 6.35 : 5.28 : 4.51 = 1 : \frac{1}{2} : \frac{1}{3} : \frac{1}{4} : \frac{1}{5} : \frac{1}{6} : \frac{1}{7}$$

The first seven observed values are in the simple ratio characteristic to a lamellar structure. Therefore, it can be estimated that this liquid crystal phase is a liquid crystal phase having a lamellar structure.

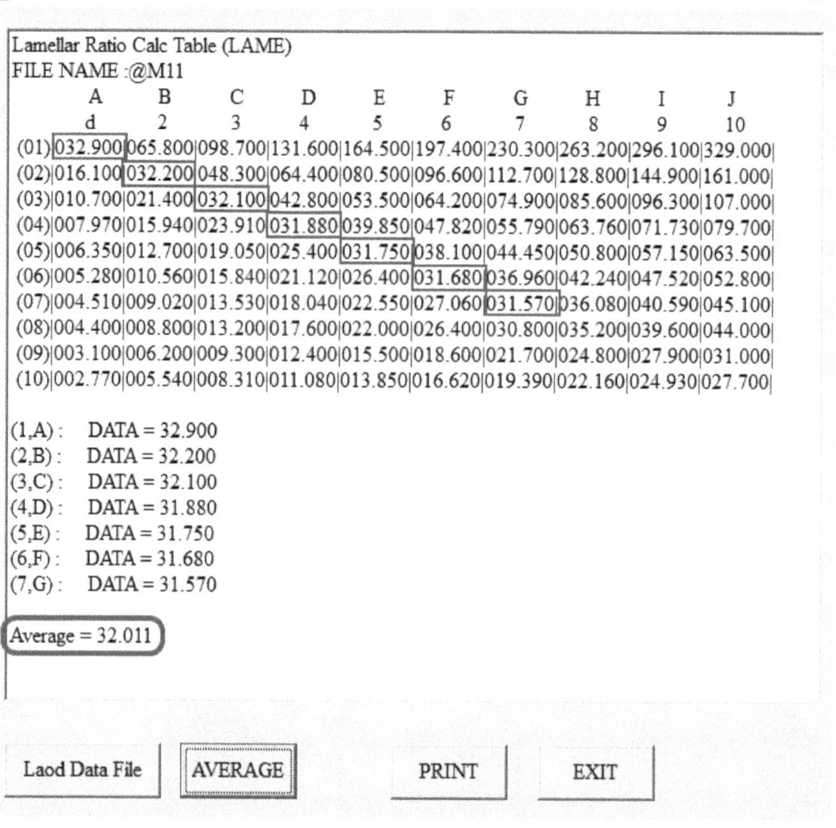

Lamellar Ratio Calc Table (LAME)

Lamellar Ratio Calc Table (LAME)
FILE NAME :@M11

	A	B	C	D	E	F	G	H	I	J
	d	2	3	4	5	6	7	8	9	10
(01)	032.900	065.800	098.700	131.600	164.500	197.400	230.300	263.200	296.100	329.000
(02)	016.100	032.200	048.300	064.400	080.500	096.600	112.700	128.800	144.900	161.000
(03)	010.700	021.400	032.100	042.800	053.500	064.200	074.900	085.600	096.300	107.000
(04)	007.970	015.940	023.910	031.880	039.850	047.820	055.790	063.760	071.730	079.700
(05)	006.350	012.700	019.050	025.400	031.750	038.100	044.450	050.800	057.150	063.500
(06)	005.280	010.560	015.840	021.120	026.400	031.680	036.960	042.240	047.520	052.800
(07)	004.510	009.020	013.530	018.040	022.550	027.060	031.570	036.080	040.590	045.100
(08)	004.400	008.800	013.200	017.600	022.000	026.400	030.800	035.200	039.600	044.000
(09)	003.100	006.200	009.300	012.400	015.500	018.600	021.700	024.800	027.900	031.000
(10)	002.770	005.540	008.310	011.080	013.850	016.620	019.390	022.160	024.930	027.700

(1,A) : DATA = 32.900
(2,B) : DATA = 32.200
(3,C) : DATA = 32.100
(4,D) : DATA = 31.880
(5,E) : DATA = 31.750
(6,F) : DATA = 31.680
(7,G) : DATA = 31.570

Average = 32.011

Laod Data File | AVERAGE | PRINT | EXIT

Step (3) To obtain the lamellar thickness, you can use the subprogram [Lame] in the initial screen [Application Menu]. Press the [Lame] button and then click the [Load Data File] button. Select the target file @M11 to download. When you further use the [AVERAGE] subprogram, you can obtain the average interlayer distance c = 32.011 =32.0 Å.

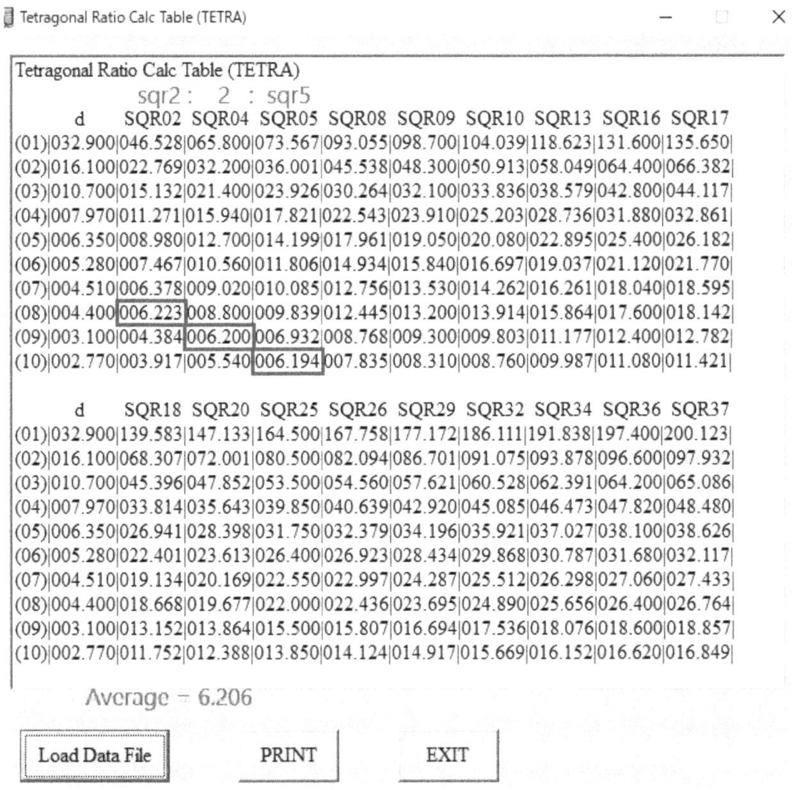

Tetragonal Ratio Calc Table (TETRA)

Tetragonal Ratio Calc Table (TETRA)

sqr2 : 2 : sqr5

	d	SQR02	SQR04	SQR05	SQR08	SQR09	SQR10	SQR13	SQR16	SQR17
(01)	032.900	046.528	065.800	073.567	093.055	098.700	104.039	118.623	131.600	135.650
(02)	016.100	022.769	032.200	036.001	045.538	048.300	050.913	058.049	064.400	066.382
(03)	010.700	015.132	021.400	023.926	030.264	032.100	033.836	038.579	042.800	044.117
(04)	007.970	011.271	015.940	017.821	022.543	023.910	025.203	028.736	031.880	032.861
(05)	006.350	008.980	012.700	014.199	017.961	019.050	020.080	022.895	025.400	026.182
(06)	005.280	007.467	010.560	011.806	014.934	015.840	016.697	019.037	021.120	021.770
(07)	004.510	006.378	009.020	010.085	012.756	013.530	014.262	016.261	018.040	018.595
(08)	004.400	006.223	008.800	009.839	012.445	013.200	013.914	015.864	017.600	018.142
(09)	003.100	004.384	006.200	006.932	008.768	009.300	009.803	011.177	012.400	012.782
(10)	002.770	003.917	005.540	006.194	007.835	008.310	008.760	009.987	011.080	011.421

	d	SQR18	SQR20	SQR25	SQR26	SQR29	SQR32	SQR34	SQR36	SQR37
(01)	032.900	139.583	147.133	164.500	167.758	177.172	186.111	191.838	197.400	200.123
(02)	016.100	068.307	072.001	080.500	082.094	086.701	091.075	093.878	096.600	097.932
(03)	010.700	045.396	047.852	053.500	054.560	057.621	060.528	062.391	064.200	065.086
(04)	007.970	033.814	035.643	039.850	040.639	042.920	045.085	046.473	047.820	048.480
(05)	006.350	026.941	028.398	031.750	032.379	034.196	035.921	037.027	038.100	038.626
(06)	005.280	022.401	023.613	026.400	026.923	028.434	029.868	030.787	031.680	032.117
(07)	004.510	019.134	020.169	022.550	022.997	024.287	025.512	026.298	027.060	027.433
(08)	004.400	018.668	019.677	022.000	022.436	023.695	024.890	025.656	026.400	026.764
(09)	003.100	013.152	013.864	015.500	015.807	016.694	017.536	018.076	018.600	018.857
(10)	002.770	011.752	012.388	013.850	014.124	014.917	015.669	016.152	016.620	016.849

Average = 6.206

Load Data File   PRINT   EXIT

The remaining observed values of 4.40 Å, 3.11 Å, and 2.78 Å are checked from the articles of "golden rule for liquid crystal structure analysis." So, the following calculations are performed:

$$\text{1D lamella } 4.40\text{ Å} \div 2 = 2.20\text{ Å} \tag{1}$$

$$\text{2D-hexagonal } 4.40\text{ Å} \div 3 = 2.54\text{ Å} \tag{2}$$

$$\text{2D-tetragonal } 4.40\text{ Å} \div 2 = 3.11\text{ Å} \tag{3}$$

See the observed values of M11 liquid crystal phase in in the above Tetragonal Ratio Calc Table (the leftmost column). Considering from the first article and Equation (1), the calculated value of 2.20 Å does not match the observed value. Next, considering from the second article and Equation (2), the calculated value of 2.54 Å does not match the observed value. Therefore, it does not have a 2D-hexagonal lattice. Further, considering from the third article and Equation (3), the calculated value of 3.11 Å just matches the observed value of 3.11 Å. Therefore, this smectic liquid crystal phase is considered to have a 2D-tetragonal lattice.

Step (4) Therefore, press sequentially the [No(N)] button for the question of "Do you calculate the lattice constant?" → [X] in the right up of inset screen → [EXIT], to return to the initial screen [Application Menu]. Press the [Tetra] button. After pressed the button, a spreadsheet appears as shown in the left. As can be seen from this table, the values in red squares are almost constant with the average value of 6.206 = 6.21 Å. Thus, it is confirmed that this phase has a 2D-tetragonal lattice.

Step (5) Press the [EXIT] button to return to the initial screen [Application Menu]. Then, press sequentially the buttons, [Recip] → [Read File] → select @M11.rcp → [Open(O)] → [FINISH] → [No(N)] → Do you calculate the lattice constant? [Yes(Y)]. Accordingly, a following screen appears.

Step (6) When appeared this screen, press the [FINISH] button.

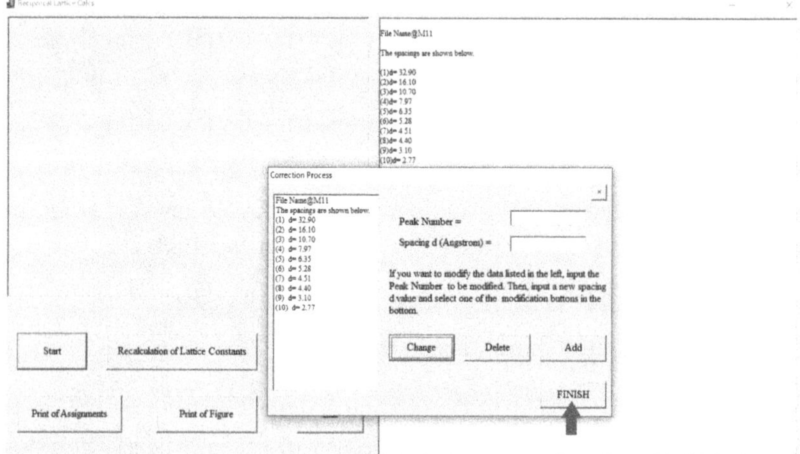

Step (7) When appeared this screen, choose the [2D-tetragonal] button.

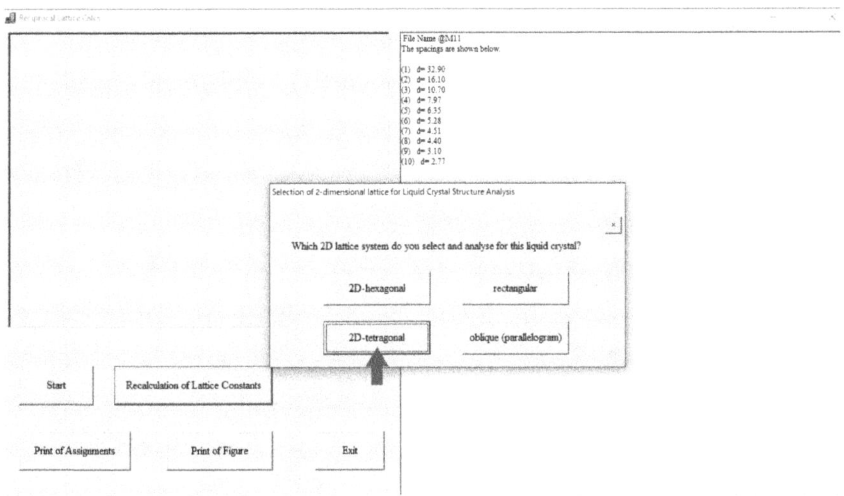

Step (8) Choose Peak No. 8 as the standard peak. Then, input (**11**) as the (hk) value of this peak.

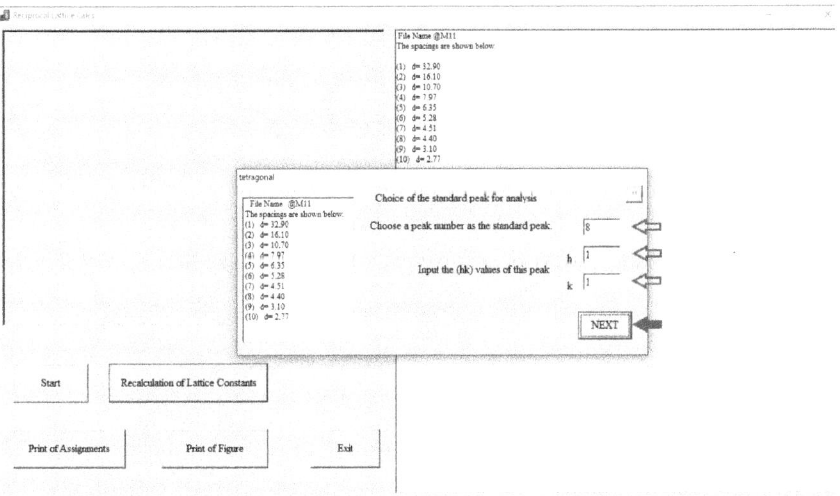

Step (9) Input the numbers of division, h and k.

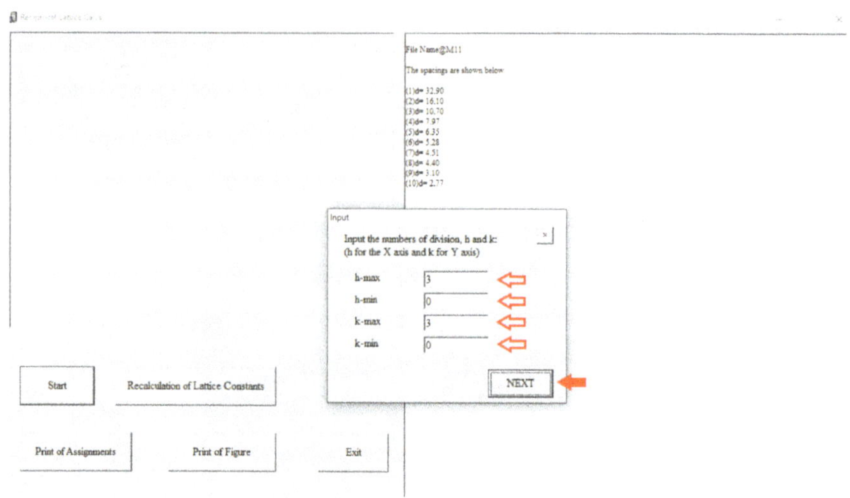

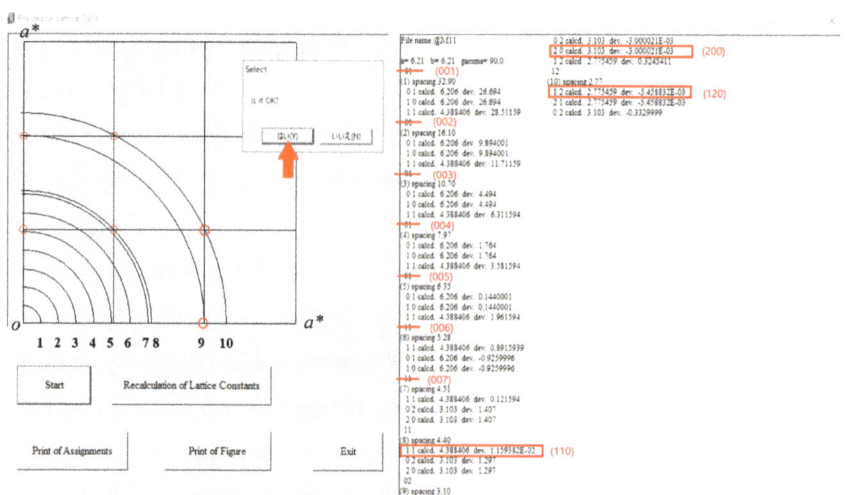

As can be seen from the figure and table, the reflections of peaks Nos. 8∼10 are in good agreement with the intersection points marked with small red circles. Therefore, this liquid crystal phase can be identified as a smectic T (S_T) phase.

Step (10) If it is OK, press the [Yes(Y)] button.

Step (11) When you press the [Yes(Y)] button in the above (10) step, a following screen appears for verification by Z value calculation. For the question [Do you carry out Z calculation?], click the [Yes(Y)] button.

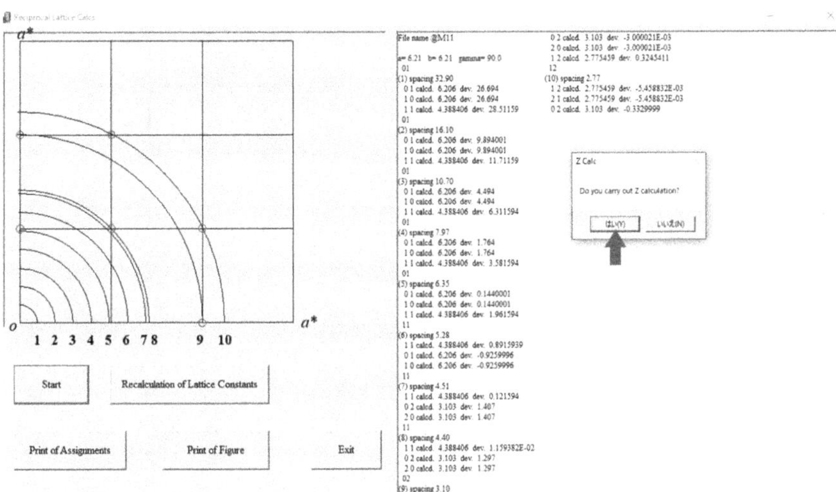

Step (12) Input the molecular weight, an assumed density and the c value in the inset table.

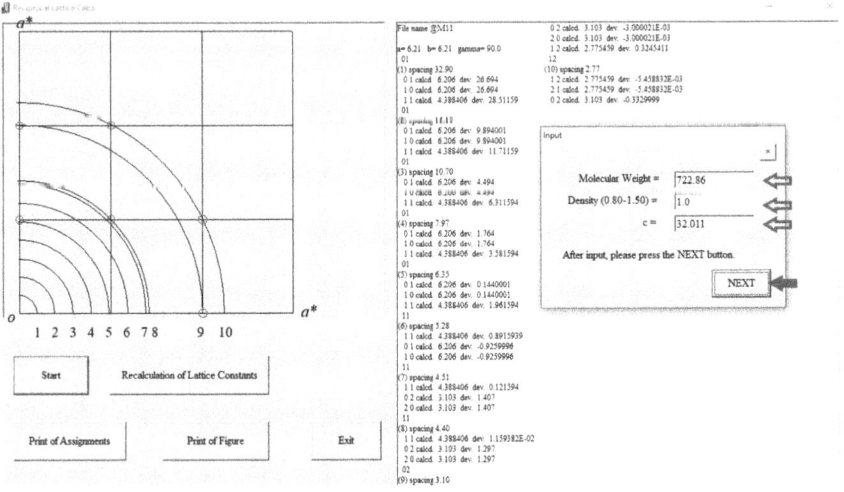

Accordingly, you can obtain Z = 1.03 ≅ 1.0, as shown in this table. The calculated Z value is an integer 1, so that this identification of smectic T (S_T) is consistently correct.

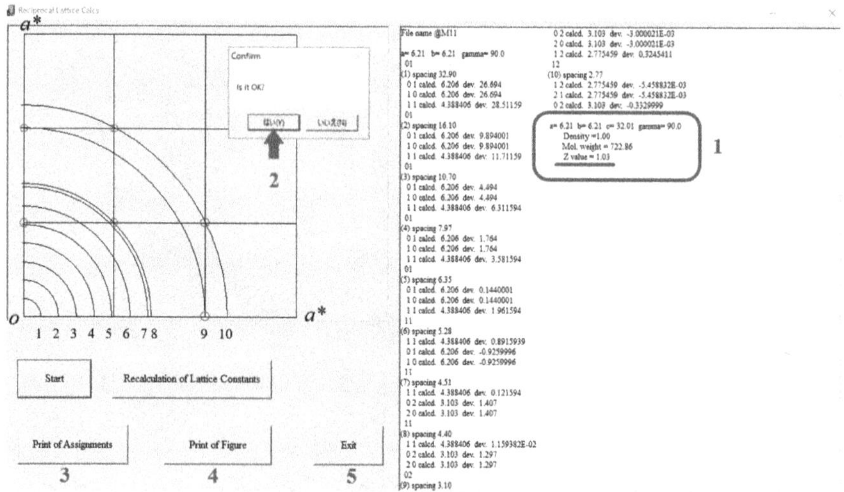

Step (13) If you are OK for the confirmation, press the [Yes(Y)] button. If necessary, press the [Print of Assignments] and [Print of Figure] buttons. Finally, press the [EXIT] button. The analysis results are summarized and tabulated as follows.

Table 14. X-ray data of @**M11** and the indexation of the reflections.

Entry No.	M11		
Reference	p76, $(C_6)_2$DABCO-Br_2		
Mw	Mw=722.86		
Temperature	120°C		
Mesophase	S_T		
Lattice constants	a = 6.21, b = 6.21, c = 32.0 Z = 1.0 for ρ = 1.0		
Peak No.	$d_{obs.}$	$d_{calcd.}$	(*hkl*)
1	32.9	32.9	(0 0 1)
2	16.1	16.0	(0 0 2)
3	10.7	10.7	(0 0 3)
4	7.97	8.00	(0 0 4)
5	6.35	6.40	(0 0 5)
6	5.28	5.34	(0 0 6)
7	4.51	4.57	(0 0 7)
8	4.40	4.39	(1 1 0)
9	3.10	3.10	(2 0 0)
10	2.77	2.78	(1 2 0)

Section 3. Structure Analysis of Quite Special Liquid Crystal Phases

11 How to Analyse @(113)Yelamaggad Add.3_$Col_{ob.o}$

X-ray Structural Analysis of Col_{ob} Liquid Crystal Line Phase by Using a New Method of "Flexible Lattice Method"

Introduction

Regarding the X-ray liquid crystal structure analysis of the Col_{ob} phase, I did not mention it at all in the previous book *Physics and Chemistry of Molecular Assemblies*. This is because it cannot be solved with paper, pencil, and calculator alone, and cannot be solved without a dedicated computer program. Accordingly, in our laboratory, we have developed a special subprogram called "**Flexible Lattice Method**" in the "Bunseki-kun program", which enables us structural analysis of this $\mathrm{Col_{ob}}$ phase. This new method will be explained here for the first time in focusing the structural analysis of a representative $\mathrm{Col_{ob}}$ liquid crystalline phase.

In Chapter 3, Problem 9 of the previous book, 110 X-ray data were posted. None of them contained a $\mathrm{Col_{ob}}$ phase. Therefore, in this book are added the following four $\mathrm{Col_{ob}}$ phase X-ray data, [(111)Komatsu Add. 1 = M8], [(112)Watanabe Add. 2], [(113)Yelamaggad Add. 3], [(114) Added Pelzl Add. 4]. Among them, [(113) Yelamaggad Add. 3] is taken up here as a representative example, and the author will explain the X-ray structure analysis method of the $\mathrm{Col_{ob}}$ liquid crystal phase by using the subprogram "Flexible Lattice Method".

11.1 Download the X-ray diffraction data

Step (1) In the "Application Menu" screen of Bunseki-kun, press sequentially the buttons, [Recip] → [Strat] → [Read File], in order to download the X-ray data of @(113) Yelamaggad Add.3.

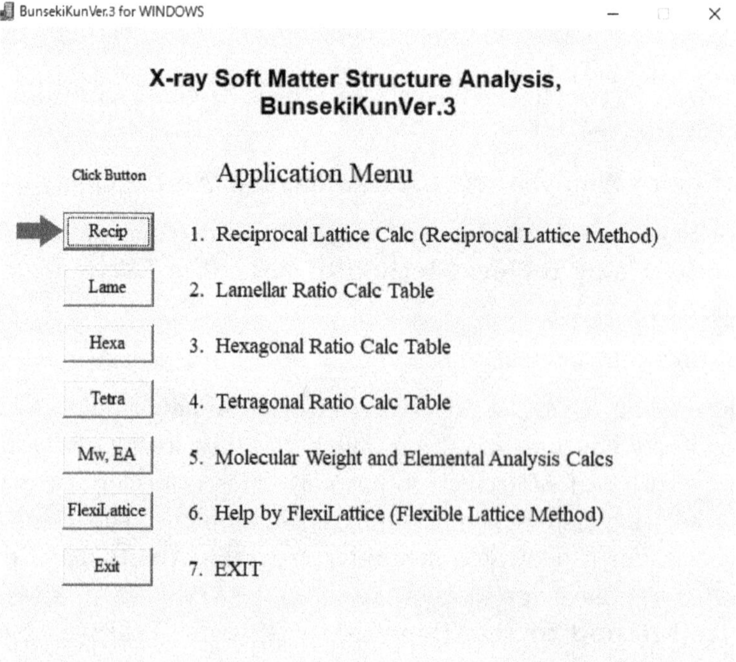

Table 15. X-ray data of @(113)Yelamaggad Add. 3.

Peak No.	$d_{obs.}$
1	**32.91**
2	**30.77**
3	**26.13**
4	**16.41**
5	**12.36**
6	**10.92**
7	**4.59**
8	**3.38**

11.2 Estimation of the dimensionality of this phase from the golden rule of liquid crystal structure analysis

Article ①: The half value of Peak No.1 can be obtained as 32.91 Å ÷ 2 = 16.455. It matches the value of 16.41 of peak No.4 within the margin of error. However, but Peaks Nos.2 and 3 cannot be explained from Article 1. Therefore, this phase is not a simple one-dimensional lamellar one.

Article ②: $32.91 \div \sqrt{3} = 19.00$, which value cannot be observed. Therefore, it is not a 2D hexagonal phase.

Article ③: $32.91 \div \sqrt{2} = 23.27$, which value cannot be observed. Therefore, it is not a 2D tetragonal phase.

Article ④: Accordingly, there remains a possibility of rectangular phase. In the rectangular phase, there should be two series of lamellar ratios in the observed values.

$$32.91 : 16.41 : 10.97 = 1 : 1/2 : 1/3.$$

Thus, you can find one series of lamellar ratios, but another series of lamellar ratios cannot be found. Therefore, it is neither a rectangular phase.

From our estimation described above, this phase has none of the lamellar, 2D-hexagonal, 2D-tetragonal and rectangular ones.

The only remaining possibility is the oblique phase (Col_{ob}).

This oblique phase cannot be analysed even by "**Reciprocal Lattice Method**" described above. This is because the angle is fixed at 90° or 60° and cannot be changed in this method.

Therefore, we have developed another analysis method named as "**Flexible Lattice Method**". This method can easily solve the oblique phase.

11.3 Structure analysis of the oblique columnar (Col_{ob}) phase by FlexiLattice method

Step (2) In the screen of Step (1), press the bottoms sequentially like as [X] → [EXIT], in order to return to the initial screen "Application Menu" in Bunseki-kun.

Then select and press the [FlexiLattice] subprogram button.

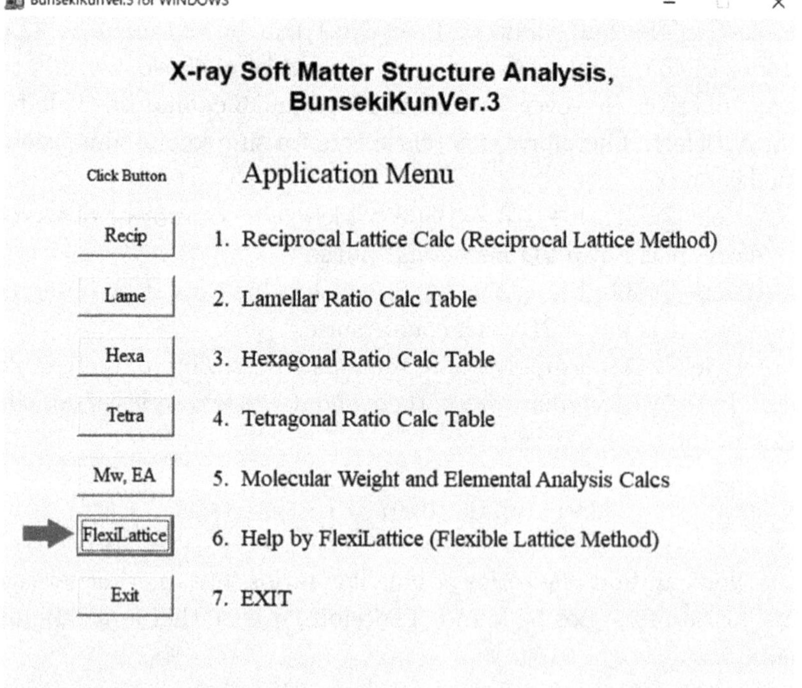

Step (3) When the screen shown in the left appears, press the bottoms sequentially: [Start] → [Read File], in order to download the X-ray data from @(113) Yelamaggad Add 3.

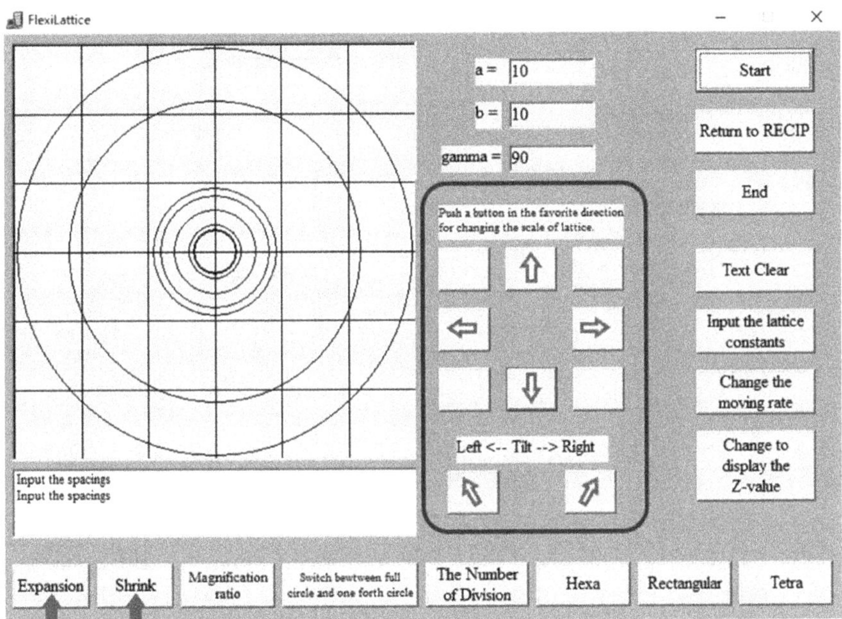

Then, press alternatively the left and right buttons and the up and down buttons indicated as red open arrows in this figure, to adjust so that the first and second peaks match to the intersection points of (10) and (01).

At this time, press alternatively the [Expansion] or [Shrink] button as to enlarge or reduce the lattice net size so that the Debye-Scherrer rings easily cross at the intersection points in the lattice.

The figure below shows the 1st and 2nd rings at lattice points (10) and (01).

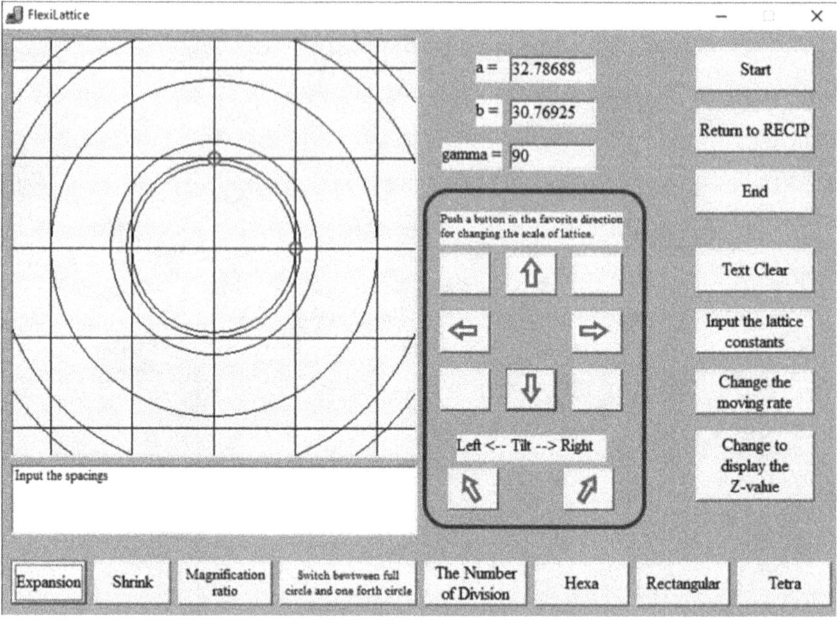

Step (4) Next, click the [Tilt] left and right buttons indicated as red open tilted arrows in this figure, to adjust so that the third peak matches the intersection point of $(1\bar{1})$. The first and second peaks, which have shifted due to tilting, are finely adjusted time to time so that they match (10) and (01).

- At this time, occasionally press the [Expansion] button in the bottom left to expand and adjust the central part. When the centre part is too expanded, press the [Shrink] button so that the whole can be seen.
- Also, if the movement speed is too fast, press the [Movement speed change] button squared in red to reduce the speed. It is recommended that the movement speed is reduced to 1/10 and fine-tuned so that all peaks are well-matched to the intersection points.

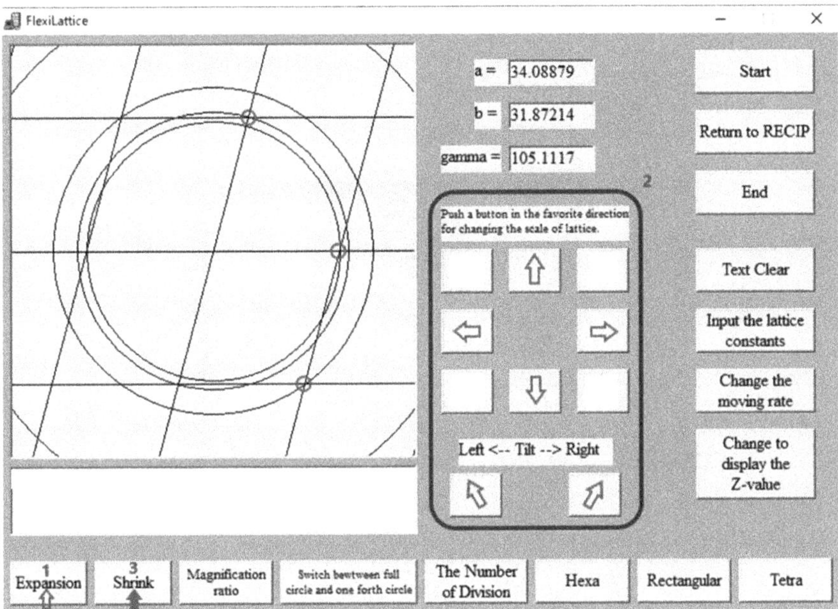

Step (5) Shrink with the Shrink button so as to see all the Debye Scherrer rings of Peak Nos. 1∼6.

Nevertheless, if you can't adjust well the screen, you can fine-tune the size of the screen by pressing the [Magnification] button in the bottom left. When the inset screen appeared, press the [Dec(rease)] and [Inc(rease)] buttons alternatively.

Then, press alternatively the left and right buttons the up and down buttons, and the tilt left and right buttons in this figure, to adjust so that all the peaks match to the intersection.

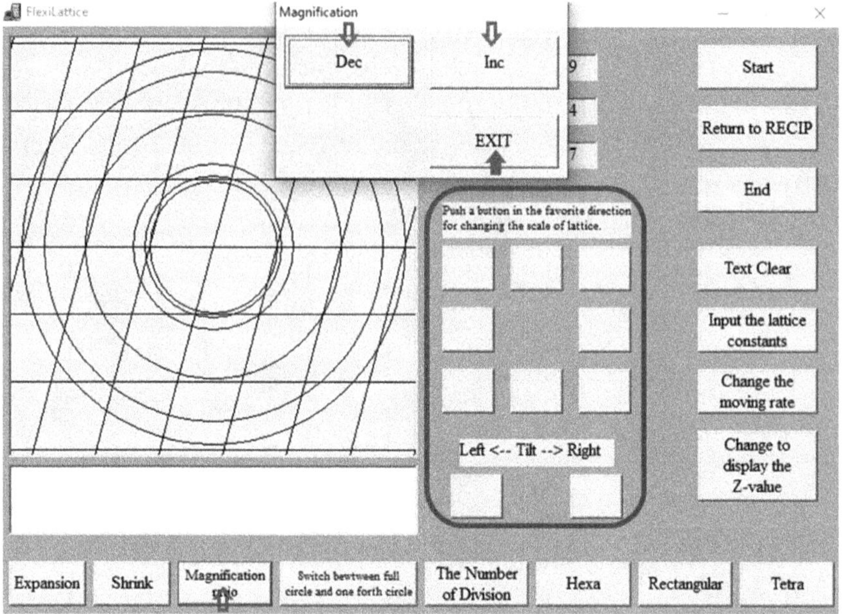

Finally, the best fit diagram is shown below.

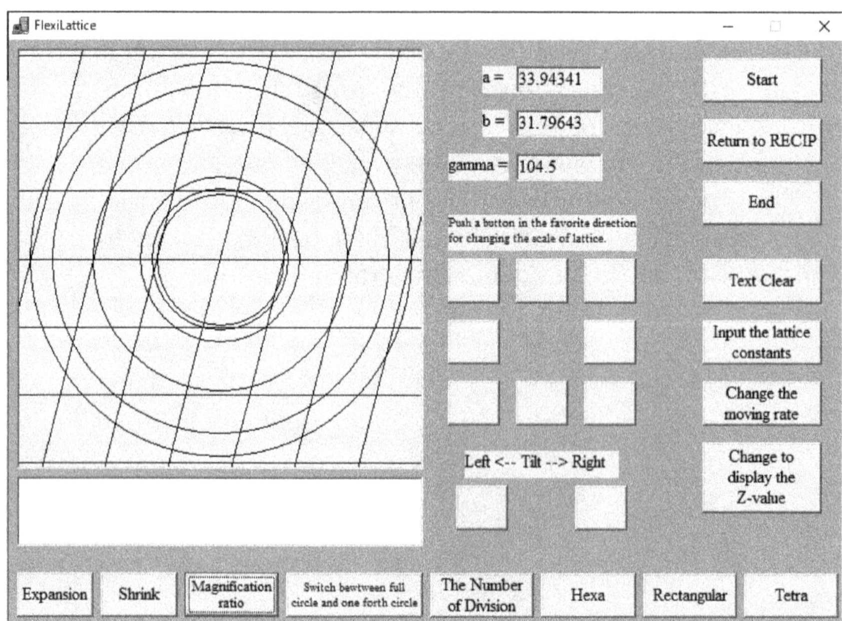

11.4 Verification by Z value calculation

The present $\mathrm{Col_{ob.o}}$ phase shows all four lattice constants, *i.e.*, a, b, c, and γ. Accordingly, you can calculate the Z value from these lattice constants. When Z is the number of molecules in the unit cell of the $\mathrm{Col_{ob.o}}$ phase, it should be Z = 1 in the $\mathrm{Col_{ob.o}}$ phase. Therefore, if the identification of this liquid crystal phase is correct, the Z value must be an integer value of 1. If it is not an integer, such as 0.5 or 1.5, the identification is wrong. To verify whether this identification is correct or not, Z-value calculation is performed as follows.

Step (6) Press the [Change to display the Z-value] button at the bottom right in this screen. When an inset screen appears, input the molecular weight, the assumed density (0.80 to 1.50), and the lattice constant c (= h: stacking distance between disks) in the blanks. Here you input the molecular weight = 1697.25 g/mol, the assumed density = 0.80 g/cm^3 and the lattice constant c = 3.38 Å. Then press the [NEXT] button.

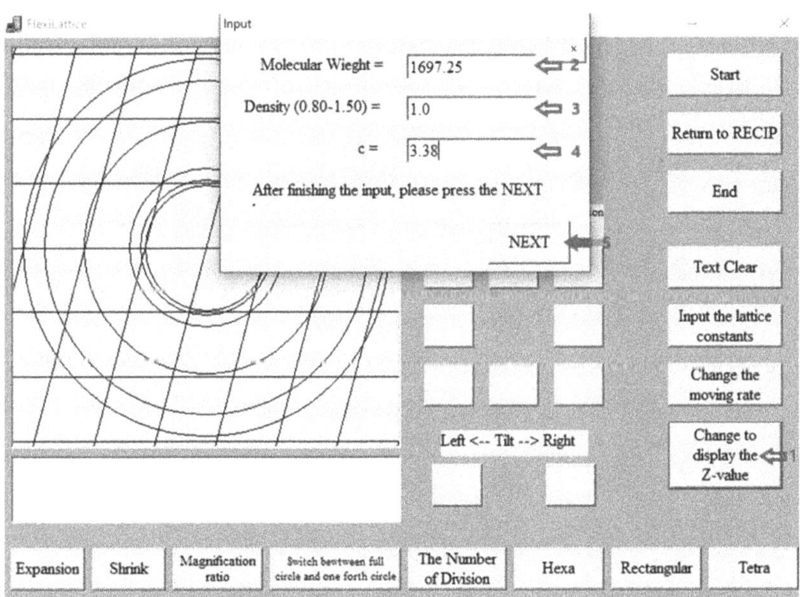

Then, the Z value (= number of molecules per unit cell) is calculated and displayed in the lower left case.

In order to show 1.00 as possible, you fine-tune the lattice constants, a, b, and γ by pressing alternately the red open arrows in the blue area. The finally obtained best-fit figure is shown in the left.

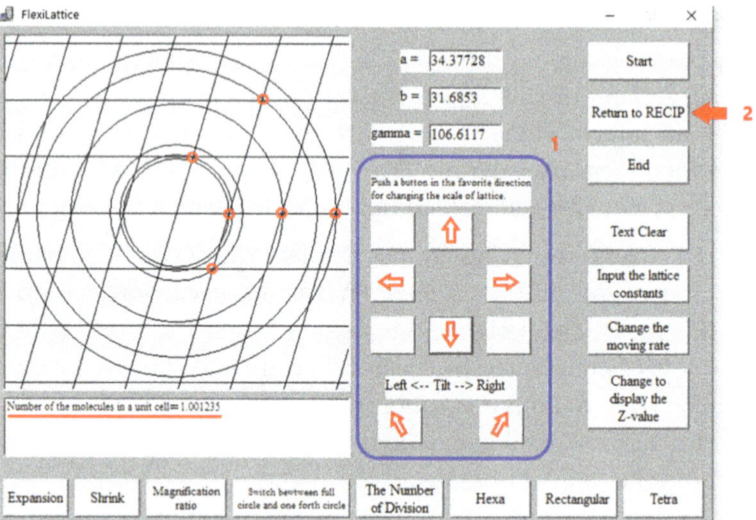

As can be seen from the lower left case, Z = 1.001235 = 1.00, which fits well with the theoretical integer 1. It supports the correctness of the present identification.

Step (7) Here, press [Return to RECIP] to obtain the calculated spacings.

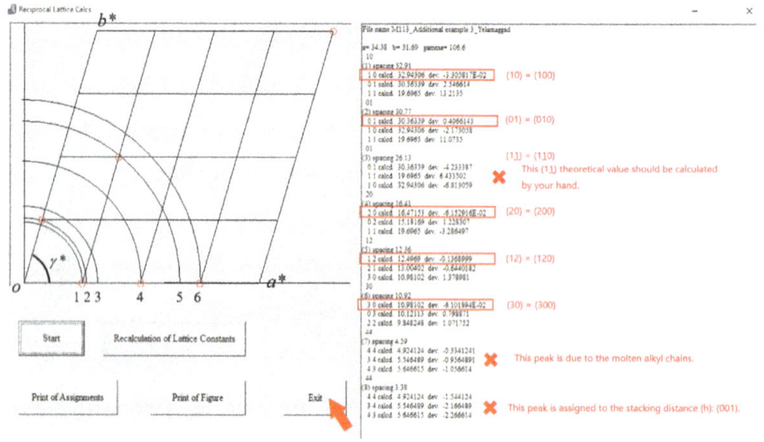

The intersection point of Peak No. 3 is located in the 3rd quadrant, so that Recip program cannot calculate the spacing value. Recip program can only calculate the spacings that match intersection points in the 1st quadrant.

Step (8) Using the lattice constants a = 34.38 Å, b = 31.69Å, and g = 104.5° obtained from the above drawing, the theoretical values d_{calcd} of the spacings can be calculated using Excel.

Table 16. Theoretical d value calculation for Col_{ob} in the sample of **@(113) Yelamaggad**.

$a = 34.38, b = 31.69, \gamma = 106.6°, h_l = 3.38$; Z = 1.00 for ρ = 0.80	Peak 1	Peak 2	Peak 3	Peak 4	Peak 5	Peak 6
a	34.38	34.38	34.38	34.38	34.38	34.38
b	31.69	31.69	31.69	31.69	31.69	31.69
γ (= gamma)	106.6	106.6	106.6	106.6	106.6	106.6
h	1	0	1	2	1	3
k	0	1	-1	0	2	0
d_{calcd}: sin(g*3.14159/180)/sqrt[h*h/a*a+k*k/b*b-(2*h*k*cos(g*3.14159/180))/(a*b)]	32.95	30.37	26.40	16.47	12.50	10.98
d_{obs}	32.91	30.77	26.13	16.41	12.36	10.92

As can be seen from this table, all the calculated values for Peak Nos. 1∼6 agree well with the observed values d_{obs}.

Peak Nos. 7 and 8 correspond to the average distance of molten alkyl chains and the stacking distance between disks, respectively. Therefore, this liquid crystal phase can be identified as the $Col_{ob.o}$ phase.

(Note) The analysis results reported by Yelamaggad *et al.* are as follows.

Table 17. Theoretical d-value calculation of Col_{ob} by Yelamaggad *et al.*

$a = 33.4, b = 31.25, \gamma = 79.9°, h_l = 3.38$; Z = 1.12 for ρ = 1.00	Peak 1	Peak 2	Peak 3	Peak 4	Peak 5	Peak 6
a	33.40	33.40	33.40	33.40	33.40	33.40
b	31.25	31.25	31.25	31.25	31.25	31.25
γ (= gamma)	79.9	79.9	79.9	79.9	79.9	79.9
h	1	0	1	2	2	3
k	0	1	1	0	2	0
d_{calcd}: sin(g*3.14159/180)/sqrt[h*h/a*a+k*k/b*b-(2*h*k*cos(g*3.14159/180))/(a*b)]	32.88	30.77	24.73	16.44	12.37	10.96
d_{obs}	32.91	30.77	26.13	16.41	12.36	10.92

x

x

There are two problems with this analysis. (1) γ should be bigger than 90° because of the crystallographic definition in a monoclinic system tilted between the a-axis and the c-axis: it is usually defined as $\alpha = \beta = 90°$ and $\gamma \geqq 90°$. Also, (2) the calculated value d_{calcd} of Peak No. 3 deviates greatly from the measured observed value d_{obs}.

Therefore, the results of the above present analysis are recommended.

The analysis results are summarized and tabulated as follows.

Table 18. X-ray data of **(113)Yelamaggad Add.3** and the indexation of the reflections

Entry No.	◎ (113)Yelamaggad Add. 3		
Reference	TLT-6 Yelamaggad		
Mw	Mw=1697.25		
Temperature	160°C		
Mesophase	$Col_{ob.o}$		
Lattice constants	$a = 34.38$, $b = 31.69$, $\gamma = 106.6°$, $h_1 = 3.38$; Z=1.00 for $\rho = 0.80$		
Peak No.	$d_{obs.}$	$d_{calcd.}$	(hkl)
1	32.91	32.94	(1 0 0)
2	30.77	30.36	(0 1 0)
3	26.13	26.40	(1 $\underline{1}$ 0)
4	16.41	16.47	(2 0 0)
5	12.36	12.50	(1 2 0)
6	10.92	10.98	(3 0 0)
7	4.59	–	#
8	3.38	–	h: (0 0 1)

C. V. Yelamaggad et al., *J. Org. Chem.,* **78**, 527–544(2013).

12 How to Analyse @(77)_Pseudo Hexagonal Phase

12.1 Download X-ray diffraction data

Step (1) In the "Application Menu" screen of Bunseki-kun, press sequentially the buttons, [Recip] → [Strat] → [Read File], in order to download the representative X-ray data of @77_Ban27.rcp.

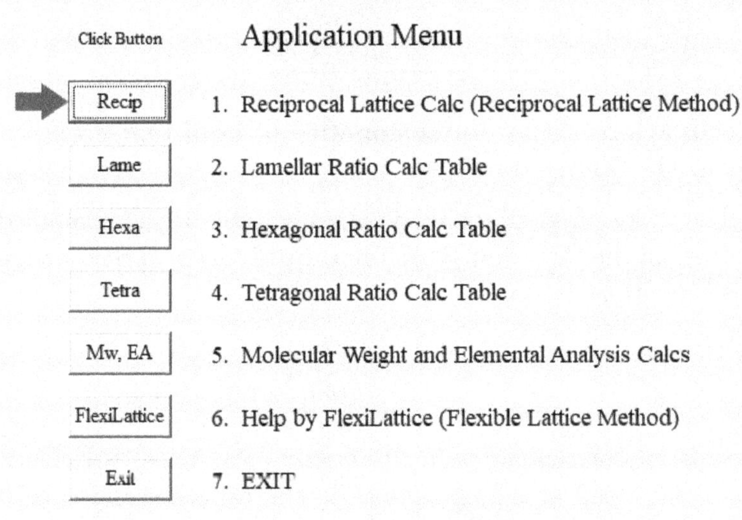

Table 19. X-ray data of @77_Ban27.

Peak No.	$d_{obs.}$
1	**28.8**
2	**22.0**
3	**17.0**
4	**14.8**
5	**11.2**
6	**9.85**
7	**9.45**
8	**8.25**
9	**7.74**
10	**7.43**
11	**5.65**
12	**ca.4.5**
13	**3.30**

12.2 Estimation of the dimensionality of liquid crystal from the X-ray data

(12.2.1) The ratios of the observed d values in Table 19 are examined from Article 2 in the "**Golden Rule for Liquid Crystal Structure Analysis**" mentioned above.

$$d_1 : d_3 : d_4 : d_5 : d_6 : \cdots = 1 : 1/\sqrt{3} : 1/2 : 1/\sqrt{7} : 1/3 : \cdots = 28.8 : 17.0 : 14.8 : 11.2 : 9.85,$$

which seems to correspond to a 2D hexagonal lattice. Strangely, however, Peak No. 2 at 22.0 Å is outside the ratio specific to this 2D hexagonal lattice.

(12.2.2) Then, the observed values are examined from Article 4 in "**Golden Rule of Liquid Crystal Structure Analysis.**"

$$28.8 : 14.8 : 9.85 \approx 1 : 1/2 : 1/3$$

$$22.0 : 11.2 : 7 : 43 \approx 1 : 1/2 : 1/3$$

Thus, you can see two series of lamellar ratios characteristic to a 2D rectangular phase.

So, you try to analyse the phase structure by assuming both 2D hexagonal and 2D rectangular lattices.

12.3 Liquid crystal structure analysis by Reciprocal Lattice Method using the standard peak

Step (2) In the screen of Step (1), press the bottoms sequentially like as [X] → [EXIT], in order to return to the initial screen "Application Menu" in Bunseki-kun.

Then select and press the [Recip]and then select the[2D-hexagonal] button.

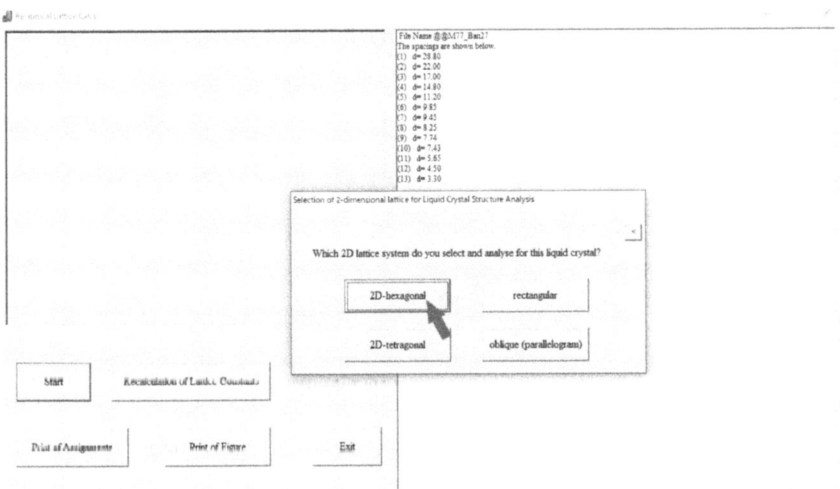

Step (3) Then, the input screen appears, so the reciprocal lattice calculation is performed by assuming that No. 3 on the slightly higher angle region, which is generally considered to have less measurement error, is the reflection from the (11) plane as the standard peak. Accordingly, input these values into the blanks indicated with red arrows. When completed, click the [NEXT] button.

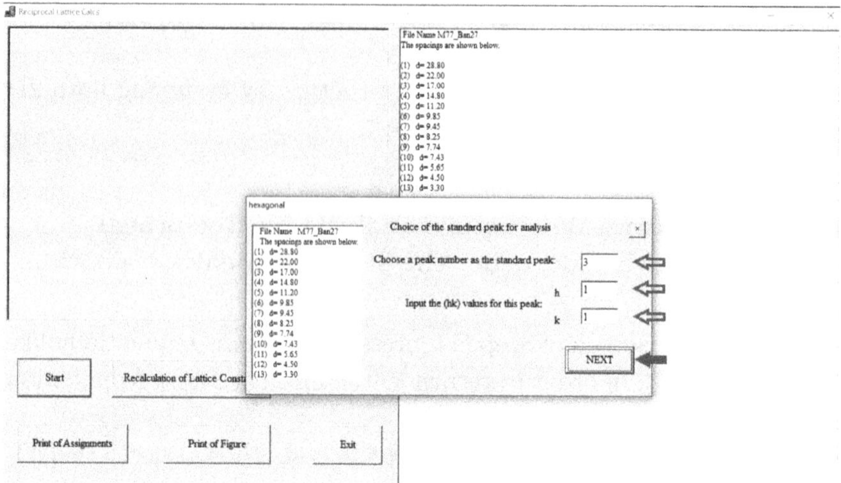

Step (4) Input the numbers, h and k, for the reciprocal lattice plane divisions. Here, you input h = 0 ∼ 8, k = 0 ∼ 8 as the number of divisions. When completed, click the [NEXT] button.

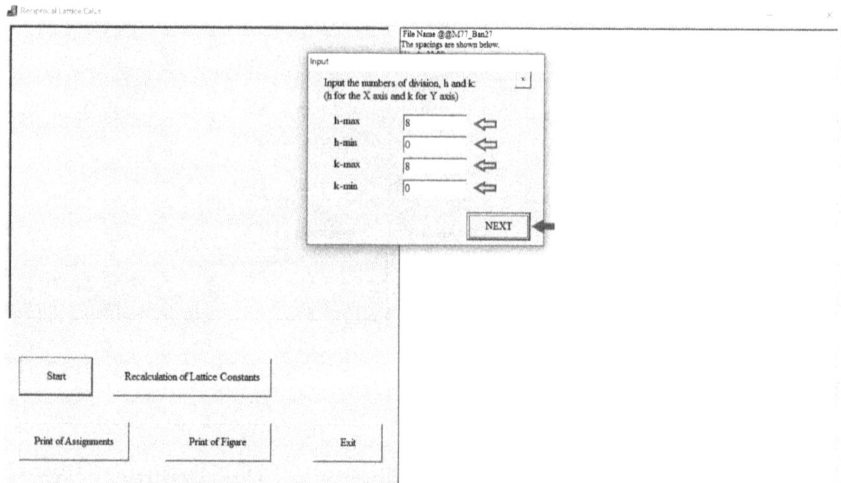

Step (5) As can be seen from the figure below, among peaks No. 1 to 11, peaks other than Nos. 2, 7, and 9 agree very well with the lattice points.

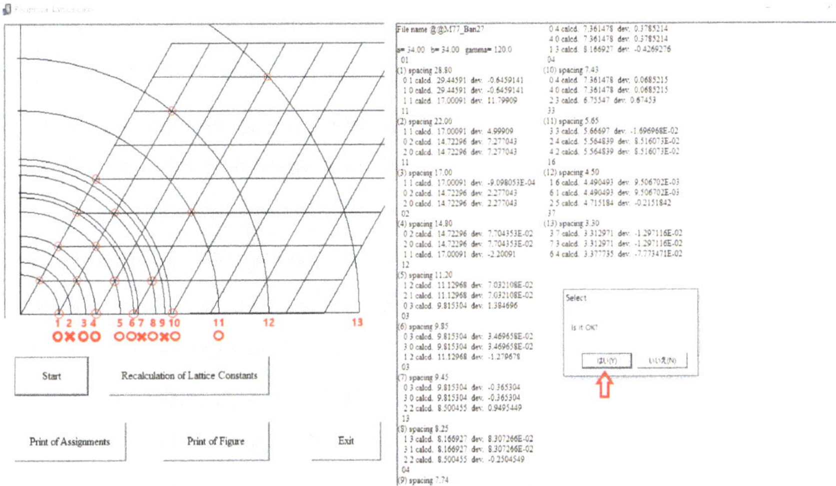

Step (6) When you press the [Yes(Y)] button in this screen, the next screen appears for verification by Z value calculation.

Step (7) When this screen appeared, click the [Yes(Y)] button for the question [Do you carry out Z calculation?]. Then, into the blanks in the appeared inset screen, enter the values of molecular weight, density and c (= stacking distance h), and then click the [NEXT] button.

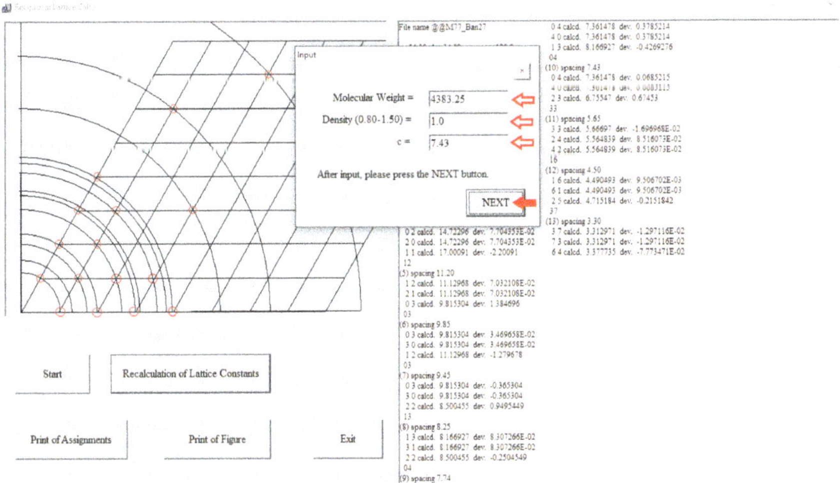

As can be seen from the table, you obtain as Z = 1.02 ≅ 1.0. Thus, the calculated Z value is integer 1, so that this identification of $\mathrm{Col}_{\mathrm{ho}}$ is consistently correct.

However, there still remains a problem that Peaks Nos. 2, 7, and 9 do not agree with the this 2D-hexagonal lattice points.

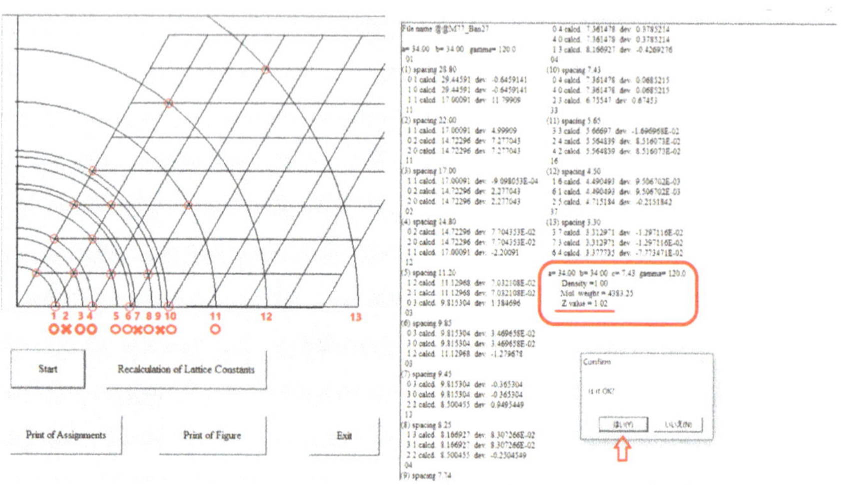

Step (8) Accordingly, press sequentially the buttons as Yes(Y) → [Recalculation of Lattice Constants] → [Yes(Y)].

Since this phase exhibits such characteristics of imperfect match to a 2D-hexagonal lattice, it is highly likely that it is a pseudo-hexagonal phase. In the case of the pseudo-hexagonal phase, the relationship a = $\sqrt{3}$b holds, so you draw a reciprocal rectangular lattice by using lattice constants a = 58.89 Å, b = 34.00 Å (a = $\sqrt{3}$b) as following manners.

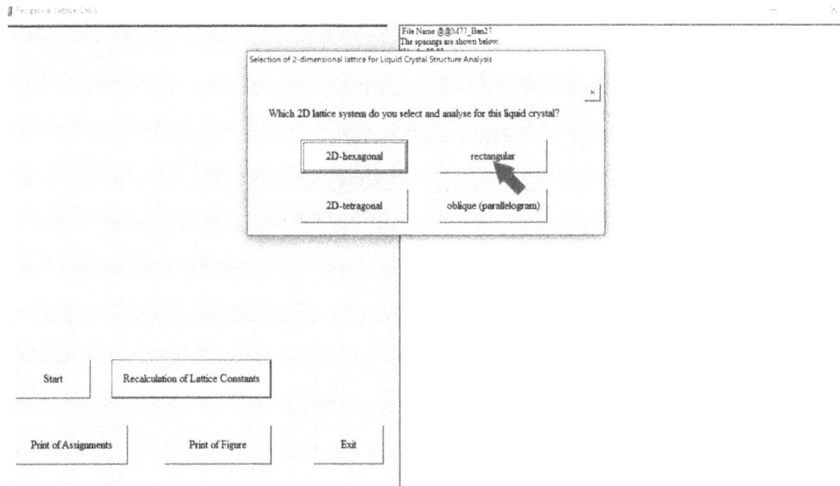

Step (9) When this screen appeared, press the [Rectangular] button.

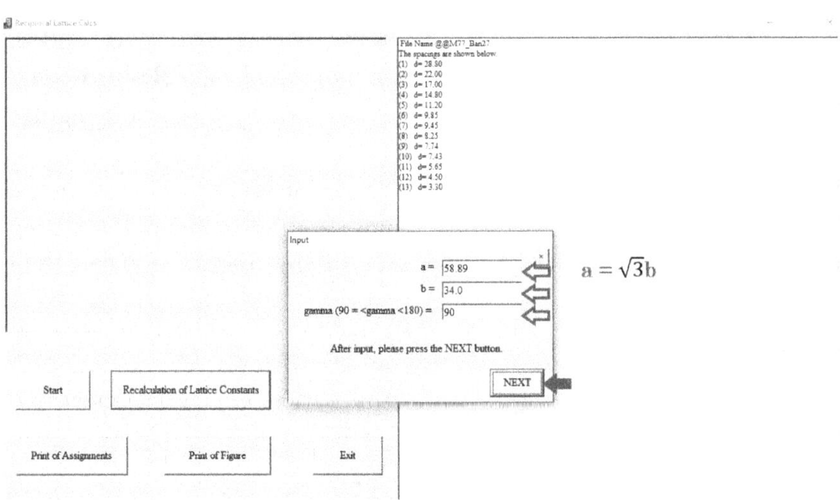

Step (10) Into the blanks appeared in the inset screen, input the lattice constants a = 58.89 Å, b = 34.00 Å ($a = \sqrt{3}b$), $\gamma = 90°$, and then press the [NEXT] button.

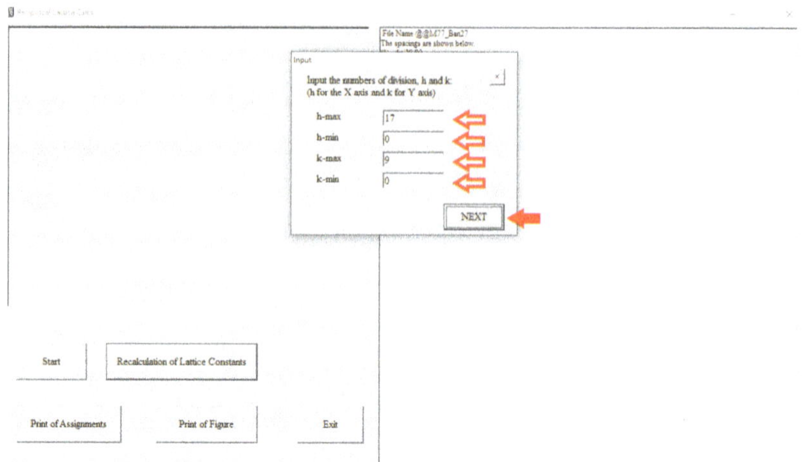

Step (11) Input the numbers of division, h = 0 ~ 17 and k = 0 ~ 9. Then press the [NEXT] button.

Step (12) Then, as shown in the figure below, Nos. 2, 6, and 8, which did not match the intersection points in the 2D-hexagonal lattice, also match very well the intersection points in this rectangular lattice.

Note: Peak No. 12 corresponds to the average distance of the melted alkyl groups, and peak No. 13 corresponds to the half stacking distance (3.30 Å($h1$) ≅ 7.43 Å($h2$) ÷ 2) of the discotic molecules in the column.

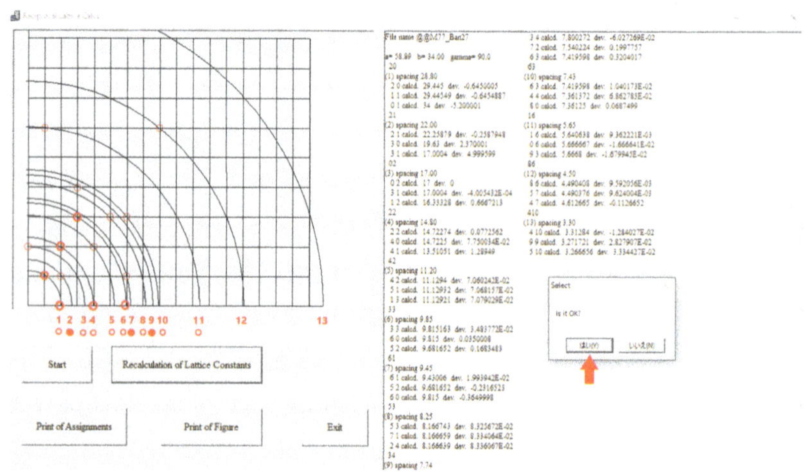

Step (13) Verification by Z value calculation.
When you press the [Yes] button in the above figure, the following screen will appear. Press [Yes] for the question of [Do you want to calculate Z?].

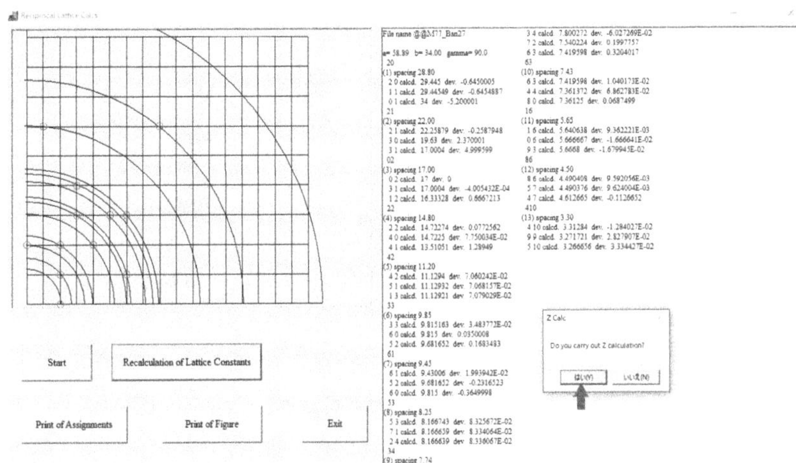

Step (14) Into the blanks appeared in the inset screen, input the molecular weight, an assumed density, and the stacking distance ($h2 = 7.43$ Å: face-to-face distance)) and press the [NEXT] button.

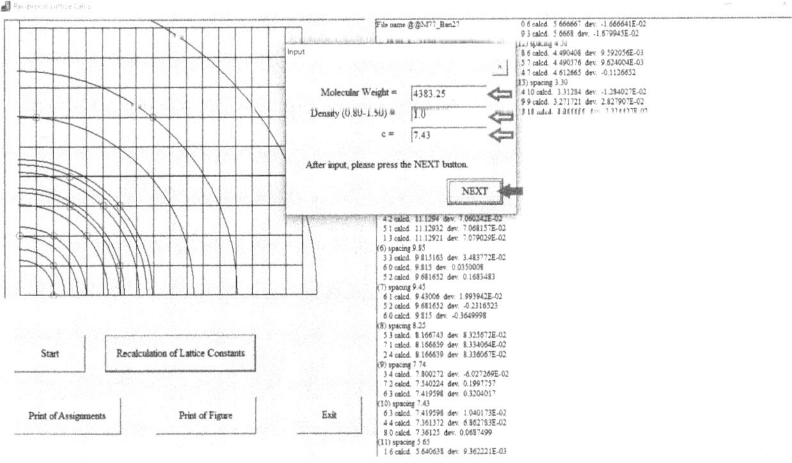

Note: This compound has a double-decker structure. Therefore, the intracolumnar stacking distance h2 shows 7.43 Å (peak No. 10) which is longer than the normal cases. This double-decker has an Eu metal sandwiched between two phthalocyanine disks. Moreover, you can also observe the half stacking distance h1 (peak No. 13), which is half of h2, because the two phthalocyanine disks can be thermally rotated to some extent. For the details, please refer to the original paper [p83. K. Ban, K. Nishizawa, K. Ohta, A. M. van de Craats, J. M. Warman, I. Yamamoto and H. Shirai, *J. Mater. Chem.*, 11, 321–331 (2001)].

Thus, you can obtain Z = 2.04 ≅ 2.0, as shown here. Thus, the calculated Z value is integer 2, so that this identification of rectangular ordered columnar (Col_{ro}) phase is consistently correct.

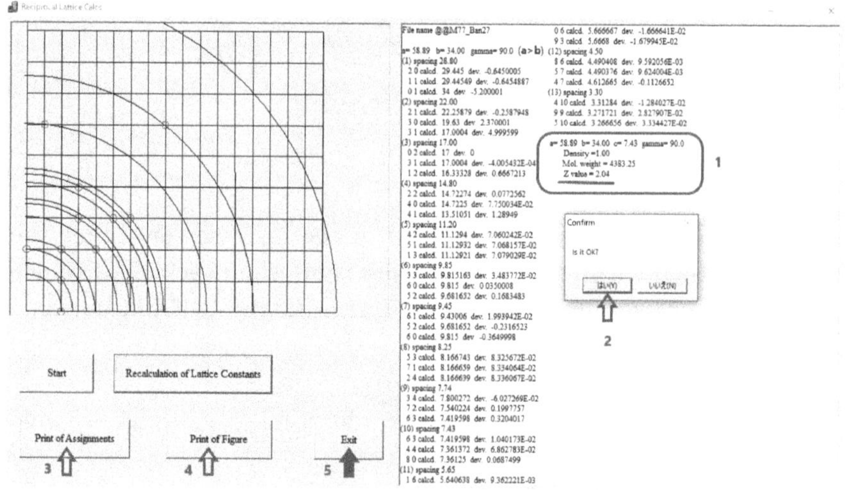

Step (15) Determine the symmetry from the extinction rules. When you furthermore consider the symmetry of lattice from the extinction rules for the rectangular lattices summarised in Figure 3, you find that this rectangular lattice has the symmetry of $P2_1/a$.

Step (16) If you are OK for the confirmation, press the [Yes(Y)] button. If necessary, press the [Print of Assignments] and [Print of Figure] buttons. Finally, press the [EXIT] button.

Step (17) Depiction of the structure of the pseudo-hexagonal columnar phase

Hereupon, you consider the reciprocal lattice of the pseudo-hexagonal lattice at first.

As can be seen from Figure 4, this reciprocal pseudo-hexagonal lattice can be easily understood by superimposing the reciprocal lattice plane of the 2D-hexagonal lattice in Step (5) and the reciprocal lattice plane of the rectangular lattice in Step (12).

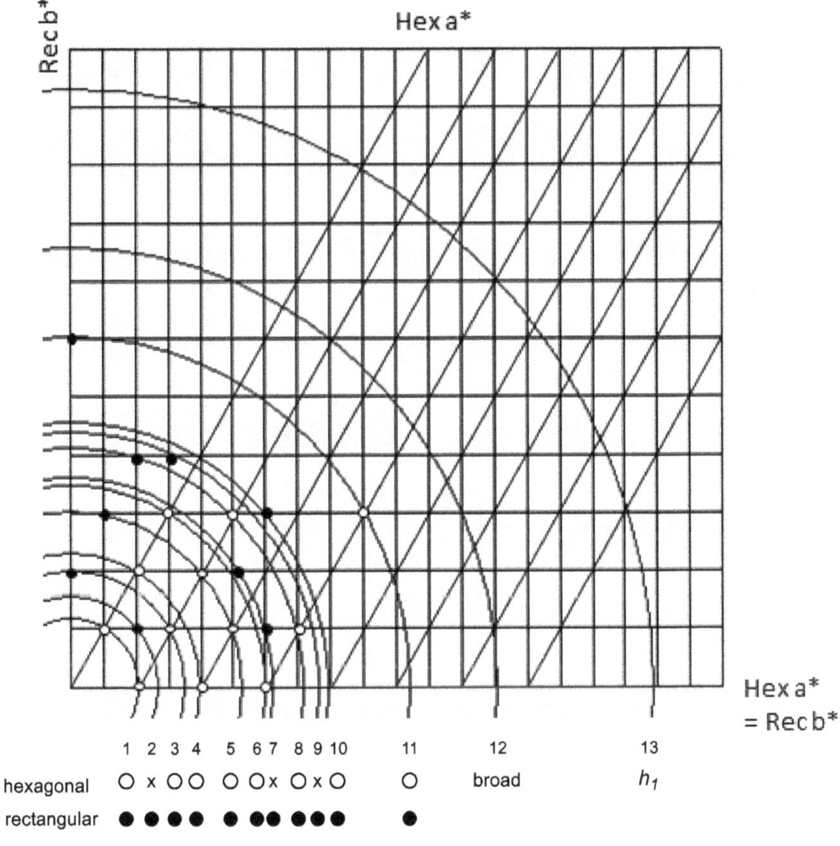

Figure 4. The reciprocal lattices of hexagonal and rectangular symmetries of $[(C_{12}S)_8Pc]_2Eu$.

Note: As can be seen from this figure, in the 2D-hexagonal lattice three peaks (Nos. 2, 7 and 9) do not match the lattice points, whereas in the rectangular lattice all the peaks match the lattice points. Moreover, in this rectangular lattice, the same peak (Peak No. 1) passes on the (11) and (20) intersection points. Normally, when you draw a rectangular reciprocal lattice, the same peak does not pass through the (11) and (20) intersection points. If so, it has a feature of a two-dimensional hexagonal lattice. In this case, you can obtain the relationship $a = \sqrt{3}b$. When you draw a reciprocal 2D-hexagonal lattice, all the peaks should fit the lattice points. Nevertheless, in the present hexagonal lattice, three peaks do not match. Such a case is a pseudo-hexagonal phase.

Three types of pseudo-hexagonal phases have been discovered so far, and they have structures of Types 1 to 3 as shown in Entry (1) to (3) of Figure 5.

- In Type 1, the whole molecules form a 2D-hexagonal lattice, whereas the central cores of the molecules form a rectangular lattice. This type of phase shows one stacking distance, which is slightly longer, at about 4.7 Å.
- In Type 2, the whole molecules form a 2D-hexagonal lattice and the central cores of the molecules form a rectangular lattice, just like Type 1. However, this type shows two different stacking distances, at about 3.5 Å and about 4.7 Å.
- In Type 3, the whole molecules form a rectangular lattice, and the central cores of the molecules form a 2D-hexagonal lattice, which is completely opposite to Type 1. This type shows one stacking distance, which is as short as about 3.5 Å.
- The bottom figure illustrates a normal structure of a general hexagonal ordered columnar liquid crystal (Col_{ho}) phase. In this structure, both the whole molecules and the central cores of the molecules form a two-dimensional hexagonal lattice. There is one type of stacking distance, which is as short as about 3.5 Å.

For the details, please refer to the following paper: p83. K. Ban, K. Nishizawa, K. Ohta, A. M. van de Craats, J. M. Warman, I. Yamamoto and H. Shirai, *J. Mater. Chem.*, **11**, 321–331(2001).

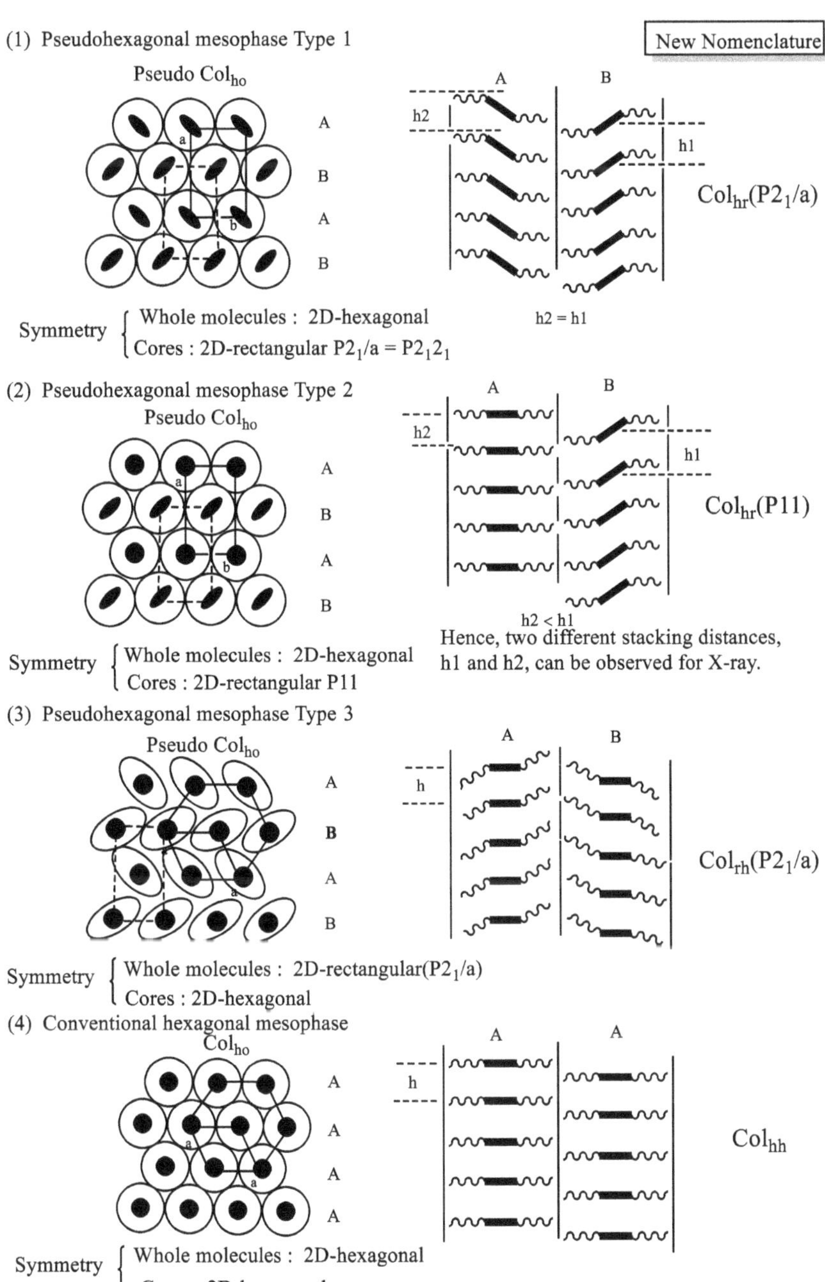

Figure 5. Relation between four kinds of 2D-hexagonal ordered columnar mesophases.

The analysis results are summarized and tabulated as follows.

Table 20. X-ray data of (77)Ban27 and the indexation of the reflections.

Entry No.	◎ (77)Ban27				
Reference	p83, $[(C_{10}S)_8Pc]_2Eu$: 1c				
Mw	Mw = 4383.25				
Temperature	45°C				
Mesophase	pseudo-Col_{ho}				
Lattice constants	Hex: a = 34.0, h_2 = 7.43, h_1 = 3.30, Z = 1.0 for ρ = 1.0; Rec: a = 58.9, b = 34.0 = hex a, h_2 = 7.43, h_1 = 3.30, Z = 2.0 for ρ = 1.0				
Peak No.	$d_{obs.}$	Hex $d_{calcd.}$	Rec $d_{calcd.}$	Hex (*hkl*)	Rec (*hkl*)
1	28.8	29.5	28.8	(1 0 0)	(1 1 0), (2 0 0)
2	22.0	–	22.3	–	(2 1 0)
3	17.0	17.0	17.0	(1 1 0)	(0 2 0), (3 1 0)
4	14.8	14.7	14.7	(2 0 0)	(2 2 0), (4 0 0)
5	11.2	11.1	11.1	(2 1 0)	(1 3 0), (4 2 0)
6	9.85	9.82	9.82	(3 0 0)	(3 3 0), (6 0 0)
7	9.45	–	9.43	–	(6 1 0)
8	8.25	8.17	8.17	(3 1 0)	(2 4 0)
9	7.74	–	7.80	–	(3 4 0)
10	7.43	7.36	7.42	(4 0 0) + h_2	(6 3 0) + h_2
11	5.65	5.65	5.67	(3 3 0)	(0 6 0)
12	ca. 4.5	–	–	#	#
13	3.30	–	–	h_1	h_1

Index

Printed in the USA
CPSIA information can be obtained
at www.ICGtesting.com
JSHW011352290923
49296JS00001B/4